THE BIODIVERSITY OF MICROORGANISMS AND INVERTEBRATES:
Its Role in Sustainable Agriculture

CASAFA Report Series

Report Series of CASAFA (Committee on the Application of Science to Agriculture, Forestry and Aquaculture) of the International Council of Scientific Unions (ICSU), published by CAB International:

1. Drought Resistance in Cereals
2. Tropical Grassy Weeds
3. Rapid Propagation of Fast-growing Woody Species

THE BIODIVERSITY OF MICROORGANISMS AND INVERTEBRATES: Its Role in Sustainable Agriculture

Proceedings of the First Workshop on the Ecological Foundations of Sustainable Agriculture (WEFSA 1) London, 26–27 July 1990

Edited by

D.L. Hawksworth

International Mycological Institute

C·A·B International

C·A·B International   Tel: Wallingford (0491) 32111
Wallingford   Telex: 847964 (COMAGG G)
Oxon OX10 8DE   Telecom Gold/Dialcom: 84: CAU001
UK   Fax: (0491) 33508

A catalogue entry for this book is available from the British Library.
ISBN 0 85198 722 2

Typeset by Alden Multimedia Ltd
Printed and bound in the UK by Redwood Press Ltd, Melksham

Contents

Organizing Committee

Dr M.S. Swaminathan FRS (*Workshop Chairman*)
Professor D.J. Greenland (*Convenor*)
Dr M.H. Arnold (CASAFA)
Sir Leslie Fowden FRS
Professor D.L. Hawksworth
Dr J.D. Holloway
Dr P. Kapoor (CSC)
Dr J.K. Waage
Mrs J.E. Boddrell (*Secretary*)

Sponsoring Organizations

CAB International (CABI)
Committee on the Application of Science to Agriculture, Forestry and Aquaculture (CASAFA) of the International Council of Scientific Unions (ICSU)
Commonwealth Science Council (CSC)
Third World Academy of Sciences (TWAS)

Contributors and Participants

[The names of contributors who were not able to attend the meeting are indicated by an asterisk (*).]

Dr M.A. Altieri; *Division of Biological Control, University of California, Berkeley, 1050 San Pablo Avenue, Albany, California 94706, USA.*

Dr M.H. Arnold; *Hamlec, 4 Shelford Road, Whittlesford, Cambridgeshire CB2 4PG, UK.*

Dr A. Baker; *Entomology Department, The Natural History Museum, Cromwell Road, London SW7 5BD, UK.*

Dr F.W.G. Baker; *Committee on the Application of Science to Agriculture, Forestry and Aquaculture, La Combe de Sauve, 26110 Venterol, France.*

Dr R.J. Baker; *Research and Development, Ministry of Agriculture, Hope, Kingston 6, Jamaica.*

Dr M. Bazin; *Biosphere Sciences (Environmental Biotechnology), King's College, Campden Hill Road, London W8 7AN, UK.*

Mr A.J. Bennett; *Overseas Development Administration, Eland House, Stag Place, London SW1E 5DH, UK.*

Professor J.E. Beringer; *Department of Botany, University of Bristol, Woodland Road, Bristol BS8 1UG, UK.*

Dr S. Blackmore; *Botany Department, The Natural History Museum, Cromwell Road, London SW7 5BD, UK.*

Mrs J.E. Boddrell; *CAB International, Wallingford, Oxfordshire OX10 8DE, UK.*

Mr I. Boler; *International Institute of Entomology, 56 Queen's Gate, London SW7 5JR, UK.*

Dr R.G. Booth; *International Institute of Entomology, 56 Queen's Gate, London SW7 5JR, UK.*

Dr J. Boussienguet; *Laboratoire de Zoologie et de Lutte Biologique, Programme National de Lutte Biologique, BP 1886, Libreville, Gabon, West Africa.*

Dr J. Bridge; *International Institute of Parasitology, 395A Hatfield Road, St Albans, ·Hertfordshire AL4 0XU, UK.*

Dr P.D. Bridge; *International Mycological Institute, Ferry Lane, Kew, Surrey TW9 3AF, UK.*

Professor A.T. Bull; *Biological Laboratory, University of Kent, Canterbury, Kent CT2 7NJ, UK.*

Dr R.G. Burns; *Biological Laboratory, University of Kent, Canterbury, Kent CT2 7NJ, UK.*

Mr A.J. Chalmers; *Redesdale Experimental Husbandry Farm, Rochester, Otterburn, Newcastle upon Tyne NE19 1SB, UK.*

Professor M.F. Claridge; *School of Pure and Applied Biology, University of Wales College of Cardiff, P.O. Box 915, Cardiff CF1 3TL, UK.*

Professor G. Conway; *The Ford Foundation, 55 Lodi Estate, India International Center, New Delhi 110003, India.*

Dr J.C. Davies; *Overseas Development Administration, Eland House, Stag Place, London SW1E 5DH, UK.*

Dr P. de Groot; *Commonwealth Science Council, Commonwealth Secretariat, Marlborough House, Pall Mall, London SW1Y 5HX, UK.*

Mr R. Dellere; *Technical Division, Technical Centre for Agricultural & Rural Coopera- tion, Postbus 380, 6700 A J Wageningen, The Netherlands.*

Dr J. Dhiman; *Department of Microbiology, Horticulture Research International, Littlehampton, West Sussex BN17 6LP, UK.*

Dr L.F. Elliott; *USDA-ARS-PWA, National Forage Seed Production, Research Center, Oregon State University, 3450 SW Campus Way, Corvallis, Oregon 97331- 7102, USA.*

Mr R.J. Fenner; *Department of Research and Specialist Services, Ministry of Lands, Agriculture and Rural Settlement, P.O. Box 8108, Causeway, Fifth Street Extension, Harare, Zimbabwe.*

Sir Leslie Fowden; *Rothamsted Experimental Station, Harpenden, Herts AL5 2JQ, UK.*

Dr F. Fox; *Xenova Ltd, 545 Ipswich Road, Slough, Berkshire SL1 4EQ, UK.*

Dr D.J. Galloway; *Botany Department, The Natural History Museum, Cromwell Road, London SW7 5DB, UK.*

Dr R. Gamez; *National Biodiversity Institute, 3100 Santo Domingo, Heredia, Costa Rica.*

Dr I. Gauld; *Entomology Department, The Natural History Museum, Cromwell Road, London SW7 5DB, UK.*

Mr C. George; *Caribbean Agricultural Research and Development Institute, University of the West Indies, St Augustine Campus, St Augustine, Trinidad.*

Dr I. Grant; *Natural Resources Institute, Chatham Maritime, Chatham, Kent ME4 4TB, UK.*

Mr L.C. Grant; *Department of Agriculture, Dunbars, Antigua.*

Professor D.J. Greenland; *CAB International, Wallingford, Oxfordshire OX10 8DE, UK.*

Dr D.A. Griffiths; *Acarology Consultants Ltd, 38 Alexandra Road, Ash, Nr Alderhurst, Hampshire GU12 6PH, UK.*

Dr M. Hadley; *United Nations Educational Scientific and Cultural Organisation, 7 Place de Fontenoy, 74700 Paris, France.*

Dr I. Haines; *Overseas Development Administration, Eland House, Stag Place, London SW1E 5DH, UK.*

Dr M. Hall; *Entomology Department, The Natural History Museum, Cromwell Road, London SW7 5BD, UK.*

Mr P.M. Hammond; *Entomology Department, The Natural History Museum, Cromwell Road, London SW7 5DB, UK.*

Dr K.M. Harris; *International Institute of Entomology, 56 Queen's Gate, London SW7 5JR, UK.*

Dr P.J. Harris; *Department of Soil Science, University of Reading, London Road, Reading RG1 5AQ, UK.*

Dato Dr Mohd Yusof bin Hashim; *Malaysian Agricultural Research, Institute, G.P.O. Box 12301, Kuala Lumpur, Malaysia.*

Dato Dr Abdul Halim Hassan; *Palm Oil Research Institute of Malaysia, P.O. Box 10620, 50720 Kuala Lumpur, Malaysia.*

Professor D.L. Hawksworth; *International Mycological Institute, Ferry Lane, Kew, Surrey TW9 3AF, UK.*

*Dr P.K. Hayes; *Department of Botany, University of Bristol, Woodland Road, Bristol BS8 1UG, UK.*

*Dr K.L. Heong; *International Rice Research Institute, P.O. Box 933, Manila, Philippines.*

Mr D.L. Heydon; *CAB International, Wallingford, Oxfordshire OX10 8DE, UK.*

Professor J.M. Hirst; *The Cottage, Butcombe, Blagdon, Bristol BS18 6XQ, UK.*

Mr D. Hollis; *Entomology Department, The Natural History Museum, Cromwell Road, London SW7 5DB, UK.*

Dr J.D. Holloway; *International Institute of Entomology, 56 Queen's Gate, London SW7 5JR, UK.*

Dr J.H. Hulse; *1628 Featherston Drive, Ottawa, Ontario K1H 6P2, Canada.*

Mr A.C. Jackson; *Technical Centre for Agricultural and Rural Cooperation, Postbus 380, 6700 A J Wageningen, The Netherlands.*

Dr K.W. Jayasena; *Agricultural Research Centre, Angunukolapelessa, Sri Lanka.*

Dr S. Jayaraj; *Tamil Nadu Agricultural University, Coimbatore 641003, India.*

Dr M.J. Jeger; *Natural Resources Institute, Central Avenue, Chatham Maritime, Chatham, Kent ME4 4TB, UK.*

Dr A. Jones; *International Institute of Parasitology, 395A Hatfield Road, St Albans, Hertfordshire AL4 0XU, UK.*

Mr C.M. Kang'e; *Kenya High Commission, 45 Portland Place, London W1N 4AS, UK.*

Dr P.M. Kapoor; *Commonwealth Science Council, Commonwealth Secretariat, Marlborough House, Pall Mall, London SW1Y 5HX, UK.*

Dr D.D. Kaufman; *Soil Microbial Systems Laboratory, Room 108, Bg 318, Beltsville Agricultural Research Center East, Beltsville, Maryland 20705, USA.*

Dr L.F. Khalil; *International Institute of Parasitology, 395A Hatfield Road, St Albans, Hertfordshire AL4 0XU, UK.*

Mr A.H. Kishimba; *Ministry of Agriculture & Livestock Development, P.O. Box 9071, Dar es Salaam, Tanzania.*

Dr R. Lal; *Department of Agronomy, The Ohio State University, Room 202, 2021 Coffey Road, Columbus, Ohio 43210-1086, USA.*

Dr G.M. Lallmahomed; *Ministry of Agriculture and Natural Resources, Reduit, Mauritius.*

Dr J. LaSalle; *International Institute of Entomology, 56 Queen's Gate, London SW7 5JR, UK.*

Dr Z. Lawrence; *International Mycological Institute, Ferry Lane, Kew, Surrey TW9 3AF, UK.*

*Dr C.M. Lazarus; *Department of Botany, University of Bristol, Woodland Road, Bristol BS8 1UG, UK.*

Dr K. Lee; *CSIRO Division of Soils, Private Mail Bag No. 2, Glen Osmond, South Australia 5064, Australia.*

Dr J.K. Leslie; *Division of Plant Industry, G.P.O. Box 46, Brisbane 4001, Queensland, Australia.*

Dr D. Lister; *CAB International, Wallingford, Oxfordshire OX10 8DE, UK.*

Professor J.M. Lynch; *Horticulture Research International, Worthing Road, Littlehampton, West Sussex BN17 6LP, UK.*

Mrs M. Mais; *Bodles Agricultural Research Station (Southern Region), Box 555, Spanish Town P.O., St Catherine, Jamaica.*

Lord Duncan McNair; *House of Lords, Parliament Square, London SW1, UK.*

Mr D. Mentz; *CAB International, Wallingford, Oxfordshire OX10 8DE, UK.*

Mr K.H. Mohamed; *CAB International Regional Office for Asia, Level 3, Block E, Kompleks Pejabat Damansara, Jalan Dungun, Damansara Heights, 50490 Kuala Lumpur, Malaysia.*

Dr M.Y.B.H. Mohiddin; *Department of Agriculture, Brunei Darussalam 2059.*

Dr A.I. Mohyuddin; *International Institute of Biological Control, PARC-IIBC Station, P.O. Box 8, Rawalpindi, Pakistan.*

Dr M.G. Morris; *Institute of Terrestrial Ecology, Furzebrook Research Station, Wareham, Dorset BH20 5AS, UK.*

Mr D.D. Moses; *CAB International Regional Office for the Caribbean & Latin America, Gordon Street, Curepe, Trinidad and Tobago.*

Professor L.A. Mound; *Entomology Department, The Natural History Museum, Cromwell Road, London SW7 5DB, UK.*

Dr R.L.J. Muller; *International Institute of Parasitology, 395A Hatfield Road, St Albans, Hertfordshire AL4 0XU, UK.*

Mr S.A.N. Muro; *Ministry of Agriculture and Livestock Development, P.O. Box 9192, Dar es Salaam, Tanzania.*

Professor L.R. Nisbet; *Xenova Ltd, 545 Ipswich Road, Slough, Berkshire SL1 4EQ, UK.*

Mr J.L. Nowland; *CAB International, Wallingford, Oxfordshire OX10 8DE, UK.*

Dr J.S. Noyes; *Entomology Department, The Natural History Museum, Cromwell Road, London SW7 5DB, UK.*

Mr R. Nyemba; *Department of Agriculture, Mount Makulu Research Station, Private Bag 7, Chilanga, Zambia.*

Dr C.P. Ogbourne; *CAB International, Wallingford, Oxfordshire OX10 8DE, UK.*

*Professor R. Olembo; *United Nations Environment Programme, P.O. Box 30552, Nairobi, Kenya.*

Dr J.F. Parr; *Dryland Agricultural Project, USDA ARS NPS, Building 005, Room 414, Beltsville Agricultural Research Center (West), Beltsville, Maryland 20705, USA.*

Dr R.R.M. Paterson; *International Mycological Institute, Ferry Lane, Kew, Surrey TW9 3AF, UK.*

Mr T.J. Perfect; *Natural Resources Institute, Central Avenue, Chatham Maritime, Chatham, Kent ME4 4TB, UK.*

Mr E.C. Philemon; *Agricultural Protection Division, Department of Agriculture and Livestock, P.O. Box 2141, Boroko, Papua New Guinea.*

Dr H.M. Platt; *Zoology Department, The Natural History Museum, Cromwell Road, London SW7 5BD, UK.*

Dr R. Plowright; *International Institute of Parasitology, 395A Hatfield Road, St Albans, Hertfordshire AL4 0XU, UK.*

Dr A. Polaszek; *International Institute of Entomology, 56 Queen's Gate, London SW7 5JR, UK.*

Dr E. Punithalingham; *International Mycological Institute, Ferry Lane, Kew, Surrey TW9 3AF, UK.*

Sir John G. Quicke; *Sherwood, Newton St Cyres, Nr Exeter, Devon EX5 5BT, UK.*

Datuk Dr Omar bin Abdul Rahman; *Prime Minister's Department, 1st Floor, Block A, Prime Minister's Department, Jalan Dato Onn, 50202 Kuala Lumpur, Malaysia.*

Ms M.S. Rainbow; *International Mycological Institute, Ferry Lane, Surrey TW9 3AF, UK.*

Dr W.D.L. Ride; *Australian National University, G.P.O. Box 4, Canberra ACT 2601, Australia.*

Dr J.M. Ritchie; *International Institute of Entomology, 56 Queen's Gate, London SW7 5JR, UK.*

Dr G.S. Robinson; *Department of Entomology, The Natural History Museum, Cromwell Road, London SW7 5BD, UK.*

Ms G. Robinson; *Bodles Agricultural Research Station, (Southern Region), Old Harbour P.O., St Catherine, Jamaica.*

Dr P.A. Roger; *International Rice Research Institute, P.O. Box 933, Manila, Philippines.*

Dr G.H.L. Rothschild; *Australian Centre for International Agricultural Research, G.P.O. Box 1571, Canberra ACT 2601, Australia.*

Dr A. Rudaz; *Swiss Federal Research Station for Agriculture, CH-3097, Uebefeld-Berne, Switzerland.*

Dr M.A. Rutherford; *International Mycological Institute, Ferry Lane, Kew, Surrey TW9 3AF, UK.*

Mrs M. Sakala; *Research Branch, Department of Agriculture, Misamfu Research Centre, P.O. Box 410055, Kasama, Zambia.*

Dr B. Samaroo; *St Clair Circle, St Clair, Port of Spain, Trinidad.*

Professor M.J. Samways; *Department of Zoology and Entomology, University of Natal, Pietermaritzburg 3200, South Africa.*

Mr G. Sandlant; *International Institute of Entomology, 56 Queen's Gate, London SW7 5JR, UK.*

Dr S.S. Sastroutomo; *Asian Plant Quarantine Centre and Training Institute, Post Bag 209, UPM Post, 43400 Serdang, Selangor, Malaysia.*

Dr P.R. Scott; *CAB International, Wallingford, Oxfordshire OX10 8DE, UK.*

Dr M.R. Siddiqi; *International Institute of Parasitology, 395A Hatfield Road, St Albans, Herts AL4 0XU, UK.*

Mr I. Simpson; *Soil Microbiology Department, International Rice Research Institute, P.O.B. 933, Manila, Philippines.*

Dr H. Somapala; *Regional Agricultural Research Centre, Department of Agriculture, Makandura, Gonawila (NWP), Sri Lanka.*

Dr W.M. Steele; *International Institute of Biological Control, Silwood Park, Buckhurst Road, Ascot, Berkshire SL5 7TA, UK.*

Professor W.D.P. Stewart; *Cabinet Office, 70 Whitehall, London SW1A 2AS, UK.*

Dr N.E. Stork; *Entomology Department, The Natural History Museum, Cromwell Road, London SW7 5DB, UK.*

Dr M.S. Swaminathan; *Centre for Research on Sustainable Agricultural and Rural Development, 11 Rathna Nagar, Teynampet, Madras 600018, Tamil Nadu, India.*

Professor R.S. Swift; *Department of Soil Science, University of Reading, London Road, Reading RG1 5AQ, UK.*

Professor I. Szabolcs; *Research Institute for Soil Science and Agricultural Chemistry of the Hungarian Academy of Science, P.O.B. 35, H-1525 Budapest, Hungary.*

Dr Mohd Senawi B. Mohd Tamin; *Malaysian Agricultural Research and Development Institute, G.P.O. Box 12301, Kuala Lumpur, Malaysia.*

*Dr P.S. Teng; *International Rice Research Institute, P.O. Box 933, Manila, Philippines.*

Dr P.B. Tinker; *Natural Environment Research Council, Polaris House, North Star Avenue, Swindon, Wiltshire SN2 1EU, UK.*

Dr D.L. Tomlinson; *Department of Agriculture and Livestock, P.O. Box 417, Konedobu, Papua New Guinea.*

Dr R. Vicencio; *Commonwealth Science Council, Commonwealth Secretariat, Marlborough House, London SW1Y 5HX, UK.*

Dr J.K. Waage; *International Institute of Biological Control, Silwood Park, Buckhurst Road, Ascot, Berkshire SL5 7TA, UK.*

Dr J.M. Waller; *International Mycological Institute, Ferry Lane, Kew, Surrey TW9 3AF, UK.*

Professor E.H. Wasfy; *Plant Pathology Department, Faculty of Agriculture, Alexandria University, Alexandria, Egypt.*

Dr R. Watkins; *CAB International, Wallingford, Oxfordshire OX10 8DE, UK.*

Dr G. Watson; *International Institute of Entomology, 56 Queen's Gate, London SW7 5JR, UK.*

Dr I.M. White; *International Institute of Entomology, 56 Queen's Gate, London SW7 5JR, UK.*

Dr M. Williams; *Bureau of Rural Resources, G.P.O. Box 858, Canberra ACT 2601, Australia.*

Dr M.A.J. Williams; *International Mycological Institute, Ferry Lane, Kew, Surrey TW9 3AF, UK.*

Dr P. Williams; *Entomology Department, The Natural History Museum, Cromwell Road, London SW7 5BD, UK.*

Dr M.R. Wilson; *International Institute of Entomology, 56 Queen's Gate, London SW7 5JR, UK.*

Mr I. Woiwod; *Entomology and Nematology Department, Rothamsted Experimental Station, Harpenden, Hertfordshire AL5 2JQ, UK.*

Dr T.G. Wood; *Entomology Department, Natural Resources Institute, Central Avenue, Chatham Maritime, Chatham, Kent ME4 4TB, UK.*

Mr Chin Fook Yuen; *Animal Feed Section, Department of Veterinary Services, Tingkat 8 & 9, Block A, Exchange Square off Semantan Road, Kuala Lumpur, Malaysia.*

Professor A.H. Zakri; *Faculty of Life Sciences, Universiti Kebangsaan Malaysia, 43600 UKM-Bangi, Selangor, Malaysia.*

Foreword

This Workshop was prompted by the need to maintain increasing agricultural production in the face of deleterious environmental changes, both natural and brought about by humans. The threats to biological diversity are particularly severe, on both the local and global scale. Much attention has been focused on readily observed changes in the diversity of plants and larger mammals. Changes in the much more numerous invertebrates and microorganisms could be of much more profound importance.

The greatest pool of genetic diversity lies among the invertebrates and microorganisms of forests and grasslands. These organisms play a vital role in maintaining and enhancing soil fertility, and hence making agricultural practices sustainable. There is ample evidence that the diversity in the biological populations can be dangerously narrowed by unwise agricultural development, and may obviate the sustainability of that development. However, much still needs to be known about biodiversity among microorganisms and invertebrates. Recent estimates of the numbers of insect species have ranged from 2 million to over 30 million – and only about 1 million have so far been described.

The Workshop was therefore arranged to address these issues, to review what is known of biodiversity among invertebrates and microorganisms, to assess how agricultural and other changes are affecting it now, and may affect it in the future, and to make recommendations about what actions need to be taken by scientists and governments to improve knowledge and determine future actions.

The review papers and discussions covered the fundamental importance of biodiversity among microorganisms and invertebrates, the functions they perform, and why they are integral to natural resource management and development.

Proposals and recommendations were made which cover:

1. The role of invertebrates and microorganisms in the conservation of biological resources, and the function and management of natural and agricultural ecosystems.
2. The need to augment the numbers of people involved in biosystematics, research on the functions performed by invertebrates and microorganisms, and ecological and other disciplines pertinent to biodiversity; the number of centres of excellence for such studies, and the promulgation of relevant expertise and information.
3. The promulgation of information about the benefits to be derived from biodiversity, and the educational and training needs for the dissemination of such information.
4. The need for international co-ordination of scientific research on biodiversity, through an international biosystematics network to co-ordinate the strengthening of world-wide capabilities in understanding and managing biological diversity among invertebrates and microorganisms.

These matters are discussed further in a Statement of Findings (Chapter 23) prepared by a Working Group convened on the day after the meeting, 28 July 1990, with the express purpose of compiling the findings of the papers and discussions which had taken place in the Workshop.

It is evident from this Workshop, the first to endeavour to address this specific problem in-depth, that policy-makers and donor organizations need to provide greater support for studies on the functional importance of invertebrates and microorganisms to sustainable agriculture, as well as population changes.

In view of the success of this Workshop, which was attended by 137 invited participants from 27 countries, CAB International is considering a series of future complementary workshops focusing on other aspects of the ecological foundations of sustainable agriculture.

Professor D.J. Greenland
Director Scientific Services
CAB International

Acknowledgements

In addition to the support of the sponsoring organizations (CAB International, Committee on the Application of Science to Agriculture, Forestry and Aquaculture, Commonwealth Science Council, and the Third World Academy of Sciences), additional financial and other support from the following organizations is also gratefully acknowledged: Australian International Development Assistance Bureau (AIDAB), British Council, Ford Foundation, The Natural History Museum, Organization for Economic Cooperation and Development (OECD), Overseas Development Administration (ODA), Royal Society, Technical Centre for Agricultural and Rural Cooperation (CTA), and United Nations Development Programme (UNDP).

The Convenor and Editor wish to thank particularly Mrs June E. Boddrell and Ms M.S. Rainbow for their professional and unstinting support before and during the Workshop, and further in connection with preparing the proceedings for publication. Dr D. Lister assisted in the final drafting of the Statement of Findings, and the draft index was compiled by Ms Rainbow.

First Workshop on The Ecological Foundations of Sustainable Agriculture

Importance of Biodiversity among Invertebrates and Microorganisms

Attendees at the Reception held at The Natural History Museum on 26 July 1990

Back Row:
S.A.N. Muro, D. Hollis, N.E. Stork, I.M. White, G. Sandlant, J. LaSalle, A. Polaszek, A.C. Jackson, R. Riley, A.H. Kishimba, M. Hadley.

Second Row:
P. Hammond, M.G. Morris, A.I. Mohyuddin, A. Jones, P.R. Scott, L.F. Khalil, M.R. Siddiqi, G. Conway, D. Lister, R. Lal.

Third Row:
K. Harrap, H. Somapala, J. Dhiman, M.S. Swaminathan, G.H.L. Rothschild, J.K. Leslie, M. Williams, A.J. Chalmers, L.C. Grant, G. Robinson, M.A. Altieri, G.S. Robinson, E.C. Philemon.

Fourth Row:
Mrs Harrap, I. Szabolcs, D.L. Hawksworth, K.W. Jayasena, J.M. Lynch, S.R.J. Robbins, Saunders, R.W.J. Keay, I. Gauld, M. Fitton, R. Gamez, I. Woiwod, P. Wightman, J.K. Waage, J.D. Holloway.

Fifth Row:
J.M. Hirst, Mrs Hawksworth, L. Fowden, A.T. Bull, I. Haines, Chin Fook Yuen, L.E. Elliott, M.H. Arnold, J.E. Beringer, I. Simpson, R.L.J. Muller, I. Grant, R.G. Booth.

Sixth Row:
L.A. Mound, T.G. Wood, D.D. Kaufman, E.H. Wasfy, K.H. Mohamed, S.S. Sastroutomo, G.M. Lallmahomed, J. Boussienguet, M.F. Claridge, B.C. Sutton.

Front Row:
J. Peake, K. Lee, P.T. Warren, S. Blackmore, J.H. Hulse, M.J. Jeger, D.L. Tomlinson, J. Bridge, M.R. Wilson, F.W.G. Baker, D.J. Greenland, R.J. Fenner, Mrs Greenland, D.L. Heydon, B.M. Dilmahamood, J.S. Noyes, Mrs Mentz, D. Mentz, unidentified, T.J. Perfect, A. Baker, P.A. Roger, I. Boler, M. Mais, R.J. Baker, J.E. Boddrell, K.M. Harris.

I
Importance of Invertebrates and Microorganisms as Components of Biodiversity

One

The Importance to Sustainable Agriculture of Biodiversity among Invertebrates and Microorganisms

W.D.P. Stewart, *Cabinet Office, 70 Whitehall, London SW1A 2AS, UK*

There is major support among enlightened world leaders for proper emphasis to be accorded to nurturing and protecting the environment. Equally, there is a need to improve the quality of human life, to improve nutrition and health-care and to minimize disease.

The advanced countries of the western World have been enormously successful in generating the crops necessary for consumption at home and abroad. This success has stemmed largely from the introduction of new plant and animal varieties, the use of fertilizers and pesticides, and from hard and productive work on the part of the farmer who has harnessed new technology to advantage as it has become available. Now that such production has been satisfied the emphasis is rightly on food quality, environmental issues, and consumer concerns.

In the developing world, the priority, generally, is to increase the quantity of food produced and for it to be effectively distributed. Although increasing attention is being paid to environmental issues, protection of the environment and the sustenance of biodiversity come a poor second when famine is the order of the day.

To begin to address the dual needs to increase food production globally and to develop a sustainable environment, that is an environment in which human needs can be met indefinitely without damaging the environment, there is a need for a rational appraisal of the problems and opportunities, and an appreciation that, while no time should be lost in arriving at practical solutions, it will take time for such solutions to become effective.

Among the issues that will have to be addressed when the maintenance of biodiversity is considered in an agricultural context are the following:

First, the extent of biodiversity in the soils of intensively developed agricultural land has declined. Nevertheless, as new policies, in countries such as the UK, lead to less of the land being used for intensive agriculture,

The Biodiversity of Microorganisms and Invertebrates: Its Role in Sustainable Agriculture.
Edited by D.L. Hawksworth. © CAB International 1991.

biodiversity is likely to increase once more in land such as 'setaside' land. Equally, biodiversity is likely to increase as toxic compounds are replaced by more environmentally benign ones, and as mixed crops and rotations replace monocultures.

Second, there will be a sustained need for the transfer, within economic constraints and realities, of existing technologies from developed to developing countries. Fertilizers and other agrochemicals, which have done so much to increase food production, cannot realistically, at a time of famine, be denied on purely environmental grounds to the countries in which such famine occurs, unless more environmentally benign alternatives are available, or unless financial incentives to allow food importation are made available. Mechanisms must be secured to sustain biodiversity in such countries while solving the food production problem.

Third, rapid progress has to be made towards the development and introduction of even safer and effective agrochemicals. In this respect the introduction in recent years of safer and more specific pesticides is particularly encouraging. This trend can be expected to continue.

Fourth, with increasing attention being paid to environmental considerations, to energy conservation, and to balance of payment deficits, the role of the natural biota will become increasingly important both in developing and developed countries. Biodiversity knows no political frontiers and there is a need to harness and exploit native biota to advantage. No opportunity should be lost to assess such potential. A sustainable environment using existing biota may be the only realistic way forward in countries with poor infrastructures and the balance of payment deficits.

Fifth, I am convinced that over the next 10–30 years advances in biotechnology will be central to improving the quality of life and will become increasingly more important as a mechanism to produce a wide variety of commercially important products. It is common sense to conserve, as far as possible, the gene pool of our planet. That is different from saying that every strain and variety on earth has to be preserved. There is increasing evidence of commonality of many genes in organisms as diverse as insects and humans, and gene exchange between microorganisms is probably quite widespread. A key question is how to address the problem of *what* can and should be conserved and of *how* to conserve that which is deemed to be important. This issue has not been adequately addressed and this conference will not have realized its objective if it does not discuss this critical problem.

My conclusion is that the problem of declining biodiversity has been belatedly appreciated, that continued and essential agricultural production in the foreseeable future demands the use of the most effective methods of generating crops while severe food shortages persist, and that increased food production must go hand in hand, as far as possible, with a national strategy linked to international effort to assess the role and importance of biodiversity. In particular, consideration has to be given to:

1. Training policy related to ensuring that globally there is a cadre of well-qualified people with expert knowledge to address and assess what can and should be done and to invoke implementation strategies. The UK with its strong research base has a major role to play in training such scientists both from the UK and from abroad.

2. Training policy, related to more public awareness of the importance of biodiversity and protecting the environment, should pervade our educational base.

3. A significantly increased research effort in the basic sciences underpinning biodiversity and including ecology, predictive modelling, soil science, taxonomy, and molecular biology, will be necessary, but research also depends on wealth creation and there is a need for wealth creation and environmental protection to go hand-in-hand.

4. A global strategy which is nationally effected.

5. The 'Best Possible Environmental Option' approach developed by the Royal Commission on Environmental Pollution should be integrated into agricultural practice in the UK, and internationally, if biodiversity is to be sustained.

6. An examination of the possibility of establishing national and internationally co-ordinated bioreserves in various parts of the world.

Overall, as with other pervasive areas of scientific research, there is a finite budget and a miscellany of problems, conflicting claims and conflicting priorities. Priorities have to be set and assessed against realistic objectives. I believe that the UK should focus its attention on the underpinning of basic sciences in a rapidly changing world, that such basic science must be adequately funded to be effective, that such work must underpin applied objectives albeit long-term ones, and that through training in the UK and by international collaboration, particularly on bioreserves, a positive contribution can be made to the difficult but not intractable problem of sustaining biodiversity within realistic and achievable goals.

Two

Importance of Microorganisms and Invertebrates as Components of Biodiversity

R. Olembo, *United Nations Environment Programme (UNEP), P.O. Box 30552, Nairobi, Kenya*

ABSTRACT: Attention is given to the conservation of microbial genetic resources. Questions which need to be answered to identify the gaps and needs to be filled by culture collections are raised. The multifarious uses and importance of microorganisms in the environment and to humans are stressed. Vast numbers have yet to be recognized and have their genetic potential realized. The vast extent of the microbial gene pool poses problems for its conservation, and the important role of collections in this is described; international co-ordination and co-operation between collections is required.

The importance of the invertebrate component of biodiversity is considered in relation to pest control, and particularly biological control. The advantages of the approach are stressed, and the available methods for the use of parasites and predators as agents of biodiversity are discussed. The characteristics to be sought in natural enemies are also summarized.

Introduction

Here I would like to draw attention to a subject about which the United Nations Environment Programme (UNEP) has long been concerned: the conservation of all kinds of living microorganisms comprising the microbial gene pool, with the aim of improving the human environment through prudent utilization of the full potential of its living resources. Harnessing microorganisms results in sound environmental management, improved biotechnology, research, training, and other scientific activities.

The Biodiversity of Microorganisms and Invertebrates: Its Role in Sustainable Agriculture.
Edited by D.L. Hawksworth. © CAB International 1991.

The first formal association between UNEP and microbiologists goes back 16 years when UNEP, the organization created by the international community to implement the Plan of Action on the Human Environment agreed at the United Nations Conference on the Human Environment in Stockholm (June, 1972) organized an Expert Group meeting on the conservation of genetic resources of microorganisms and the possible contribution of microbiology to the UNEP programme.

Stockholm produced principles and recommendations emphasizing, *inter alia*, the need for conservation of all genetic resources. The Conference was the first world-wide intergovernmental meeting to declare decisively that conservation should be part of the rational use of resources, such rational use being based on appropriate scientific knowledge. And, although it pre-dated the biotechnology revolution, its recommendations made liberal references to the potential applications of microorganisms in environmental management. Eighteen years after the UN Conference on the Human Environment, it is now appropriate to take stock on where we stand and, in the light of increased demands on the role of biodiversity in improving human welfare and sustainable development, to decide and recommend appropriate strategies which could assist us in nudging the world towards its current goal of sustainable development.

Microbial genetic resources

Here we can address a number of questions. Let me start by those relevant to microbial genetic resources:

1. Can we ever hope to create a comprehensive microbial gene pool, and is this a necessary goal?
2. What is the current status of the existing microbial culture collections?
3. Does enough and appropriate information flow out of the major and small collections available around the world today?
4. Do we have reliable preservation techniques?

Answering these questions allows us to identify the gaps and needs to be filled in existing collections, identify specific strategies required to rescue endangered and unique collections world-wide, develop priority plans for all culture collections, and further our knowledge in conservation methodology. No one can claim to be able to answer these questions adequately. Thus, I shall address the generality of microbial resource conservation and the problems associated with such activities.

Microorganisms are the most numerous and ancient biotic entities which have successfully colonized every possible ecological niche on this planet. Their presence and activity is essential to the health and functioning of ecosystems and the whole environment through, among others, mineralization and recycling of important elements, photosynthesis, proper conditioning and fertility of the soil, pest and vector control, and detoxification of naturally occurring and man-made pollutants. They also play a significant role in bioproductivity, either directly *via* synthesis of food, feedstuffs, medicines, and chemicals, or indirectly by making nutrients available for other primary producers. The role they play in nature to maintain the dynamic equilibrium and integrity of the biosphere is of such magnitude that the continued existence of biological life itself is crucially dependent on the sustained microbe-mediated transformation of matter in terrestrial and aquatic ecosystems.

The extraordinary activity of microorganisms is based on their remarkable metabolic diversity and genetic adaptability. However, the fact that almost all biological processes in our environment are directly or indirectly governed by microorganisms is often overlooked, and the potential benefits of regulating, optimizing and exploiting microbial action has hardly been fully tapped. One reason is the relative youth of microbiology as an independent discipline of scientific enquiry. The organized study of microbiology is only slightly over a hundred years old, and, since its early days, microbiology research primarily and justifiably focused on microorganisms associated with human and animal disease and with food production and preservation. This interest subsequently widened in many countries and microbial technologies for the production of antibiotics, making of oral contraceptives and other medicines, management of pests and pathogens, bioleaching of metals, increasing soil fertility, generating biofuels, creating perfumes, monitoring air pollution, destroying persistent pollutants, ridding coal mines of methane, assaying of chemicals, cleaning up of oil spills, waste water treatment, and serving as tools for medical research. Microbes are well-established and becoming increasingly significant in national economies. However, those microorganisms that are well-studied and have been isolated, characterized, preserved and, sometimes, exploited by humans, constitute only a minute fraction of Earth's microbial genetic diversity. But what of the vast majority of microorganisms that are beneficial or essential to the continued functioning of the biosphere? Established collections of microorganisms with a wide scope of activities, particularly for ecological and environmental applications, are still lacking.

Vast numbers of microbial species are yet to be recognized (Hawksworth and Mound, Chapter 3) and their genetic potential realized. The use of beneficial microorganisms is far from being exhausted. Among others, the three major pressing problems facing the expanding world population – food and energy production and waste free technologies – could benefit from microbiological research. The potential of microorganisms in solving problems facing both developed and developing countries has been recently recognized, and many national, regional and international institutions have initiated relevant programmes. Many new developments can be expected in the next decade, and conservation of the genetic potential of microorganisms is essential to those developments.

Conservation of the microbial gene pool

In principle, every genotype and every constituent DNA sequence is unique, irreplaceable, and hence deserves conservation. However, microbial genetic diversity is so broad and is in such a dynamic state that any attempt to conserve it *in toto* in captivity would be unthinkable. In the case of the fungal gene pool, for example, the number of species of fungi currently maintained in culture collections throughout the world represents 17% of the 69 000 accepted species of fungi, which itself is far short of the total number of fungal species and has long been quoted as 250 000, but are now conservatively estimated at 1.5 million (Chapter 3). Despite the fact that conservation of microorganisms is far more simple and less expensive than conservation of plants and animals, one can therefore never hope to capture the full breadth of the total gene pool in culture collections. There is a cost element in conservation and also a benefit that can be attached to a given genome. Considering the financial resources required, it is unrealistic to assume that all existing microbial resources will ever be preserved. Priorities, therefore, must be established.

With the recent developments in the area of biotechnology and related techniques, such as recombinant DNA and computerized information retrieval and dissemination, it is evident that microbial resources will undoubtedly contribute increasingly to the development and improvement of agriculture, health and environmental management. The relative ease of genetic manipulation of microorganisms and the universality of the genetic code in all biological entities makes it possible to utilize microorganisms for dramatic optimization of selected plant and animal characteristics, and even for the conservation of specific genetic material of certain plants and animals through genetic engineering.

If culture collections are the cornerstone of biotechnology, without reliable and adequate collections there can be no advances in this area, and biotechnological processes and innovations could not be sustained. There is a need for further development of culture collections and for the exchange of information on cultures to make a significant advance toward sustainable development. Culture collections must be kept aware of current developments in microbiology and biotechnology so that they can focus on maintaining and providing germplasm and data that are most needed. Microbial culture collections must include new strains (and engineer new and improved ones) to represent as much of the microbial gene pool as is possible. Valuable germplasm in culture collections constitutes a resource not only for development, but also for research, training, and education. Yet conservation should be geared to sound utilization, otherwise microbial gene banks will become mere storehouses, for which no one will pay.

In many developed countries there are well-established national or international, specialized services and institutional and private culture collections which are equipped to maintain cultures with regard to phenotypic, genetic, and other traits. They act as repositories for the safe storage, preservation, and distribution of many thousands of microbial strains each year, including highly specialized and type cultures of new genera and species. Irrespective of these facilities, collections of cultures built up for specific purposes or projects are disappearing or becoming endangered just as science is learning how to exploit their genetic variability and to gain more information as to their potential benefits. Over the last few years, a number of collections in developed countries have either been threatened with closure, or have been deprived of the support necessary to function adequately. On the other hand, established service culture collections do not exist in most developing countries, and the majority of the collections in those countries centre around individual research programmes. Culture collections have not adequately developed in the majority of cases as the maintenance of culture collections is considered to be a specialized service that is time consuming, not particularly glamorous, prestigious, or directly rewarding. Many cultures are prone to loss or become endangered because of the absence of adequately trained personnel, storage facilities, funds, awareness of the value of the collections, or knowledge of existing local or university collections. Meanwhile, they cannot depend on service culture collections in developed countries for reasons of lack of information, cost, quarantine, distance, and postal and patent regulations.

Culture collections are primarily built up through isolations made from natural habitats, supplemented by type or other reference strains acquired from other collections, and there are no data on how effectively various natural habitats have been ecogeographically surveyed for microorganisms. Natural habitats frequently contains microorganisms with useful traits that would have disappeared had they not been collected and safely preserved in an unchanged form. It is true that because of their occurrence in huge numbers, very rapid reproduction, and broad genetic diversity, it is likely that specific microbial traits become extinct only under extremely harsh environmental interferences by human activities through pollution, concentrated accumulation of organic wastes, massive application of chemical fertilizers and pesticides, destruction of ecosystems, etc. Retrieval of particular microbial strains exhibiting specific traits from natural ecosystems is, however, technically almost impossible. Indeed the only way to make such a trait readily available for exploitation is by safely conserving microbial strains with desired traits in service culture collections. But without sufficient research, strains with valuable traits may be lost during isolation, preservation, or storage. Also, with the loss of microbial habitats due to the development of encroaching settlement, we can no longer rely on being able to ever reisolate and replace lost material. In this connection, it is useful to bear in mind that geographic, environmental, geological and climatic factors often make the microbial flora of one location significantly different from another, in spite of the versatility of microbial strains. The areas of possibly the greatest microbial genetic diversity are undisturbed habitats in the tropical and subtropical zones which are frequently located in the countries least well-equipped to initiate programmes necessary for the isolation, characterization, preservation, and exploitation of the available material.

Culture collections are restricted by local needs, interests, and regulations. In industry, activities are aimed at short-term commercial or research goals. Accessibility to industrial or government collections is restricted by patent regulations, commercial interests, or security measures, all of which limit the flow of information and hence the possibilities for wider exploitation.

Proper maintenance of culture collections requires convenient methods for maintaining those organisms that retain their viability in a genetically and physiologically stable form. Small culture collections without adequate preservation facilities are obliged to maintain cultures in the actively living state at the risk of an increased mutation or loss under unfavourable circumstances. Traditionally, curators kept culture collections as working collections at lower temperatures (4°C, which requires subculturing the material every 4 weeks or so). Some claim that, if the germplasm were lost in working collections, it could be re-collected, but this might be impossible, as pointed out above. In the case of fastidious groups, for example, cultures are often lost despite sincere attempts to maintain them because of either the lack of proper communication to the curators from the depositors, or the lack of basic scientific information on the physiological characteristics of the organisms. Many culture collections still operate manual documentation systems because they cannot afford the initial cost of the computerized data systems that both save curators' time and allow for specificity and rapidity in information retrieval. They would facilitate feedback on results obtained when strains are distributed as such information should be gained and added to the documentation on the collection.

Another problem is the replication of cultures within and between collections. How many cultures are replicated? There is a need to deposit cultures in replicates as a safeguard against their loss and as maintenance becomes difficult when culture holdings tend to increase.

Factors that may lead to the loss of a collection or its becoming endangered include: lack of adequate financial or manpower support, changes in the interests of individual's or institution's research programmes; retirement, transfer, or death of particular scientists associated with the collection; careless preservation, power or equipment failure; lack of understanding of the role of collections in microbiology, biotechnology, and other areas; and lack of a sound basis for microbial taxonomy.

There is consequently a clear need to identify the gaps and extent of overlap between existing collections, and to take initiatives to remedy these by international co-ordination and co-operation. There is a need for a continuing forum for collections to exchange information on the state of collections, and for the interchange of expertise between curators and the scientific community, in order to identify the most serious problems and to initiate appropriate actions for their solutions.

The importance of invertebrates in pest control

Regarding the importance of invertebrates as components of biodiversity, I will restrict myself to predators and parasites as important components of biodiversity for pest control.

The use of parasites and predators forms part of the science of pest control known as 'biological control', that is the action of parasites, predators and pathogens in maintaining other organisms at a population density lower than that which would occur in their absence. Parasites and predators for the confines of this submission are taken to imply insect parasites (parasitoids) and predators, both of which are included in the general category of biological control agents known as 'natural enemies'.

Insect parasites are often called parasitoids to distinguish them from the typical invertebrate parasites which rarely kill their hosts. However, the two terminologies are often used interchangeably. By its nature, a parasitoid requires just one host to complete its development, and with time such a host is usually killed. The adults are free-living, often feeding on honeydew, pollen and extra-floral secretion. Parasitoids therefore simply kill for their progeny, and the phrase 'protelean parasites' has consequently been used to describe them. Unlike a parasitoid, a predator kills food for itself. It requires more than one host to complete its development. Both the adult and immature forms are predatory in habit. The predatory habit is equally widespread in many insect groups, especially Hymenoptera, Hemiptera, Coleoptera, Diptera, and Neuroptera.

With regard to the ecological basis for the use of parasites and predators in pest management, the following points merit emphasis. Pest problems are population problems, a population being a functional unit of the community, which in turn is a component of the environment. The environment is therefore an important parameter in pest population dynamics. An insect becomes a pest when its population grows above an acceptable level in relation to human agricultural activities.

All living organisms are endowed with a certain maximum potential rate of increase characteristic for each species. Based on this potential rate of increase, the change in numbers of an insect population from generation to generation may be estimated. As a population further increases, there is more or less a proportional decrease in the rate of increase until the rate becomes zero and the population levels off. This is simply because as a population increases beyond a certain level, food and space become limiting, deterioration in the quality of the population itself becomes evident, it attracts more natural enemies and therefore mortality increases until the population is again brought down to that certain level which approximates to what the energy resources of its environment can support. That is the equilibrium level for that population, and is the carrying capacity for that particular environment.

Parasites and predators are mortality factors which form an important part of the biotic components which in general operate in a density dependent manner. That is, the severity of their action increases as the population increases above the equilibrium, and relaxes as the population falls below that equilibrium. A population is consequently prevented from over-exploiting its resources or from dwindling into extinction.

What methods exist for the use of parasites and predators as agents for increasing biodiversity? Many agricultural pests are insects that have been inadvertently moved from their original home to a new locality where they now become pests. For example, approximately 61% of the agricultural pests in the USA are known to be exotic in origin.

The natural enemies which keep such insects below pest status in their native situations are usually left behind in the process of accidental movements; hence a population-limiting factor is removed and they become pests. Such pests are amenable for control by the importation of parasites and predators. The importation of natural enemies has therefore been largely directed towards exotic insect pests. However, a substantial number of native pests also have been controlled by the importation of natural enemies.

A good natural enemy serves no useful purpose unless it can be successfully colonized and established in the locality to which it has been imported. This demands skill, patience and perseverance. There must be an adequate number of natural enemies for release, and they must be introduced near the appropriate stage or stages of the pest. An inherently effective natural enemy may be limited in prey population regulation by some intrinsic factors peculiar to it; these may include: (i) an inability to adapt to weather extremes; (ii) an inability to synchronize with the necessary stages of the host; and (iii) adverse effects of periodically unsuitable environmental conditions.

There are two basic approaches to biocontrol using parasites and predators: (i) mass production and periodic colonization; and (ii) planned genetic improvement. Most of the pioneer work on the periodic mass releases of natural enemies has been with species of the egg parasite *Trichogramma*. In Mexico, as well as in Colombia, there are government *Trichogramma* insectaries where thousands of *Trichogramma* insects are produced annually for use against various pests, including bollworm *Heliothis* species, and pink bollworm *Pectinophora gossypiella*. Since 1963, the losses in cotton production in Torreon, Mexico, have declined substantially each year also with a significant reduction in insecticide usage. In the USSR, where most work on mass releases of *Trichogramma* has been conducted, about 6 million acres received releases of *Trichogramma* in 1968 against more than 10 pest species, among which are armyworm, cabbage moth, corn borers, potato tuber moth, and other moths infesting grains.

Note: Professor Olembo unfortunately had to withdraw from the Workshop at short notice, but his paper was tabled and distributed to all participants.

Three

Biodiversity Databases: The Crucial Significance of Collections

D.L. Hawksworth, *International Mycological Institute, Ferry Lane, Kew, Surrey TW9 3AF, UK*
L.A. Mound, *Department of Entomology, The Natural History Museum, Cromwell Road, London SW7 5BD, UK*

ABSTRACT: Biodiversity is a crucial component of the functioning of the world's ecosystems and vital to the maintenance of their stability. The number of organisms involved, especially of fungi and insects, is almost incomprehensible, with 1.5 million and 6–80 million species respectively. In collecting and describing these, priority needs to be accorded to those of functional importance as 'keystone' species. Reference collections which are permanently preserved are a crucial component of the information transfer system of biological diversity. They are essential for identification, as vouchers for the application of names, vouchers for species used in research projects, and are also databases themselves – incorporating a massive body of information on distribution, seasonality, etc. There is a need for world initiatives to harness this information. Collections can also provide the basis of training, can exchange specimens to assist new centres being established, and are well-suited to produce a wide variety of publications at different levels. The need to strenthen expertise in the developing countries is stressed, through co-operation with facilities already present in the developed world.

The background

We are the product of biodiversification. Evolution did not progress by a straight line from microbe to *Homo sapiens*. The complex genetic system which we call 'our species' could only come into existence through the success, and the failure, of vast numbers of other genetic systems. Today we sometimes behave as if this process is now complete. Our crops are seen as

The Biodiversity of Microorganisms and Invertebrates: Its Role in Sustainable Agriculture.
Edited by D.L. Hawksworth. © CAB International 1991.

man-made products growing in a medium of soil and atmosphere which we determine, and as former Prime Minister Margaret Thatcher stated in 1989, we have assumed that the efforts of humankind would leave the fundamental equilibrium of the world's systems and atmosphere stable (Thatcher, 1989).

Such a simple, mechanistic viewpoint of our place in *Gaia* is not tenable. Biodiversification is not just the means by which *Homo sapiens* came into existence. The diversity of living things is intimately related to the function and stability of communities and ecosystems. Larger plants and animals do not operate in nature as single organisms. The existence of each depends on a variety of others, and fundamentally on the microorganisms as the basis of food pyramids (Price, 1988). The ecosystem function of biodiversity is rarely addressed, and has concerned the International Council of Scientific Unions' (ICSU) Scientific Committee on Problems of the Environment (SCOPE; di Castri & Younès, 1990). This applies not just to undisturbed ecosystems, but also to crop systems and even to the well-being of human communities with their need for recreation in gardens, parks, forests, and other open spaces. Human culture world-wide, our music, writing, painting and sculpture, is itself deeply rooted in our instinctive appreciation of biodiversity.

The present Workshop, focusing on the biodiversity of microorganisms and invertebrates in sustainable agriculture, is less about the past (i.e. our origins and those of our crops and animals) or even the future (i.e. finding new crops and improving old ones) than it is about the present (i.e. our current existence). *Homo sapiens* relies on the earth's biodiversity. Our crops are pollinated by insects, and fungi and other microorganisms are essential as mutualistic symbionts in the stems and roots of crop plants as well as in the guts of our ruminant livestock. Terrestrial ecosystems are inextricably linked together by fungi and insects. Fungi, in addition to functioning as mutualists, have major roles as natural biocontrol agents of insects, nematodes, and plants. Insects feed on fungi and plants, and are fed on by other insects as well as by bats and birds, frogs and fish, lizards and crocodiles, rats and monkeys. This becomes particularly obvious when an organism is introduced to a new area in the absence of its natural predators and parasites. The links weaken and the ecosystem is disrupted by the new pest. One reason why the maintenance of natural ecosystems is vital to future agricultural production is that those natural systems represent the source of future biocontrol agents (Morris *et al.*, 1991).

Most importantly, dead plants and animals are degraded by microorganisms and invertebrates, their stored energy and nutrients being released and made available to the next generation. Without this recycling the essential biological basis of life would collapse. We only become aware of this when something goes wrong. For example, in Australia, Norris (1974) estimated that cow-dung, which cannot be digested by Australian insects because they have evolved to feed on native marsupial dung, would cover 2.4 million hectares of pasture in a single year. Similarly, without fungi able to break down fallen trees and other plant litter (Hudson, 1986), large areas would assume the aspect of ever-growing compost heaps. Biodiversity is consequently essential, maintaining the stability of living systems around us, and ensuring that we survive the death, dung, and detritus of contemporaneous members of our own species as well as of other organisms.

The number of species

Despite our reliance on living systems the biological database on which we plan our future is remarkably sparse. About 250 000 species of seed plants are thought to exist (Wilson, 1988), but scarcely 150 of those have entered commerce. Human populations world-wide support themselves to a very large extent on four species: maize, rice, wheat, and potatoes (Plotkin, 1988, p. 107). Our knowledge of the biology of most plant species is fragmentary, and our knowledge of the biology of most animal species, apart from a few vertebrates, is effectively non-existent. More surprising to most people is the absence of any integrated list of the species of organisms which have been described since the formal start of biological nomenclature in the 1750s.

Current estimates of the number of living species for several groups are given in Table 3.1. From this it is clear that insects and fungi include the largest number of species. A third group of organisms may yet be found to join these two, the nematodes, but recognition of species among these worms remains difficult and estimates of the number of species remains suspect.

The wide variation in the estimates of the number of insect species in the world is due to our ignorance of insect biology. We do not have sufficiently reliable data yet to be sure what proportion of insects are specific to particular hosts, or are limited in their geographical or habitat range. Similar problems exist with estimates of the number of fungus species as increasing levels of complexity and new habitats are discovered progressively, with fungi living in the surface of marble, associated with bubbles of foam, inside vascular plants and liverworts, on insects bodies and in their guts, and on other fungi (Hawksworth, 1991a). But whatever the upper limits of the numbers of living species, our knowledge of their functional importance in this world remains fragmentary. Currently, less than 1 million species of insects have been described, but the life histories and immature stages are known for scarcely 1% of these. The biological diversity of the world is almost unbelievably great. We know we are dependent on it. But currently we cannot measure it satisfactorily and our estimates of the loss of biodiversity, our working capital, are therefore conjectural. On this base of ignorance we are planning our future occupation and development of planet Earth. Any business plan which included such a market

Table 3.1. World species totals, based on multiple sources, including Hawksworth (1991a); Wilson (1988); Wolf (1987); Stork (1988);* Gaston (1991),[†] Holloway and Stork (Chapter 5).[†]

Group	Described species	Estimated species
Dicot plants	170 000	
Monocot plants	50 000	
Ferns	10 000	
Mosses and liverworts	17 000	
Fish	19 000	21 000
Birds	9 000	9 100
Mammals	4 000	4 000
Reptiles and amphibians	9 000	9 500
Algae	40 000	60 000
Protozoa, etc.	30 000	100 000
Nematodes	15 000	500 000
Bacteria	3 000	30 000
Viruses	5 000	130 000
Fungi	69 000	1 500 000
Insects	800 000	2 000 000
		to
		80 000 000*
		or
		5 000 000
		to
		10 000 000[†]

evaluation would be seen as quite unacceptable by even the most insensitive investor.

How are these figures for the numbers of species in major groups derived? In part they are based on careful samples at particular sites, or even from individual trees (Holloway & Stork, Chapter 5). Not all the species collected and recognized in such samples need to have been described or named; but they must be distinguishable from each other. This is by no means a simple matter. It may be possible for a competent biologist to recognize most species of butterflies, macrofungi, lichens and even moths at a particular site, at least in Europe or parts of North America. But few individual biologists could do this in any tropical area. Even in Europe, an entomologist who can identify every species of beetle collected in an oak woodland is rare, and identification in many groups of microfungi requires critical microscopic study. Thus species lists at particular sites usually depend on the services of both experienced collectors and observers and access to a range of specialist taxonomists. With some groups of organisms scanning electron microscopy, biochemical tests, and isolation into pure cultures in the laboratory can be required.

The nematodes, also of considerable agricultural and medical significance, are equally poorly known. If 500 000 species are really present in the world as estimated by Poinar (1983), only about 3% of the world's nematodes are yet known.

Completing the inventory of the world's biota for microorganisms and invertebrates will clearly be a major task. At the current rate of progress, time and resources are not on the side of the taxonomists. In the case of the fungi, for example, at the rate of progress achieved over the last 250 years it will take another 844 years (Hawksworth, 1991*a*). Sponsorship of species descriptions as proposed by Quicke (1991) is unlikely to make any significant contribution outside mammals, birds, reptiles, fish, butterflies and moths. What can be done is to: (i) focus taxonomic resources on the groups that are of functional importance in ecosystems as 'keystone species' (Heywood, *in* Dayton, 1991), or are most likely to be of value to humans; and (ii) take steps to ensure that the time taxonomists have available for research is not wasted on diversionary activities, notably nomenclatural and historical investigations (Hawksworth, 1992; Ride, 1991), that contribute little of significance to our knowledge of the species themselves.

Recognizing the numbers of species at one site is only a first step. The basic knowledge about a group of organisms must be such that reasonable predictions can be made concerning the specific identity of similar looking organisms found at different sites. This process is more complex, and is dependent not just on competent and knowledgeable taxonomists, but also on the availability of an adequate reference collection and library.

From the initial estimate of the number of beetles or bugs at a site, and an estimate of the degree of host specificity they are likely to show as well as the number of available plant species, broad estimates can be built up of the total number of insect species likely to exist (Stork, 1988). Such methods can produce very high estimates of the number of living species, depending on the assumptions made about host-specificities, and as a result have been subject to criticism (Hodkinson and Casson, 1991).

In contrast, Gaston (1991) produced a much lower total figure by asking a wide range of insect taxonomists to estimate the total number of undescribed species likely to exist in the groups on which each specialized (Table 3.1). Ultimately our knowledge of biodiversity depends on detailed studies by taxonomists on particular groups of organisms, and the systematically arranged collections of specimens and libraries which are housed in museums, universities and research institutes. Most biologists study restricted taxonomic groups, perhaps one family or order, or specialize in a particular habitat; they are conscious of the fact that many other species exist, but only teams of taxonomists, with their broad field collecting experience, their extensive libraries and catalogues, and their rich collections of reference specimens, are in a position to convey some measure of the extent of biological diversity.

Collections and information transfer

Named reference collections are the basic tool for communicating information about biodiversity between different workers in time and space.
It may not be difficult to distinguish two species of small brown beetles
one from the other; but to recognize consistently the same two species
from a variable assemblage of 10 or 100 similar species is much more
difficult. Voucher specimens or cultures of commonly encountered species
used in investigations and experimentation are essential if workers are
to be sure that they are studying the same species from one generation to
another and at different sites, and that their results will be accurately
indexed in the international literature. Researchers in any aspect of biology
should deposit reference specimens in appropriate institutions so that
even if an organism is wrongly named, or its name changes, the work
retains its value as the subject's identity can be verified. Too much biological research, often elegant and costly, is already little more than waste
paper.

The scientific names of organisms of all groups are based on a system of
nomenclatural types, particular specimens to which those names apply. Type
collections are consequently of immense practical importance in determining
how a name is to be applied. The collections of microfungi at the International Mycological Institute, insects at The Natural History Museum, and
nematodes at the International Institute of Parasitology between them
preserve the type material for more of these organisms of agricultural importance than any other institution.

The crucial significance of collections is that they are the basis for the
application of names, the means to transfer information, and make possible
the interpretation of published data in scientific journals such that this can
be reconfirmed in further field and laboratory studies. Without voucher
specimens or cultures it is not possible to be certain that subsequent studies,
or even parallel studies elsewhere, involve the same species.

In practice, the situation is far from satisfactory. Reviewing our knowledge of arachnids in the USA, Coddington *et al.* (1990) pointed out that
the taxonomic literature consists of hundreds of uncoordinated papers
with no comprehensive identification guides nor keys to genera. Expertise in arachnids in North America is thus based on the collections and
libraries of three institutions, and training in this subject is largely an
oral tradition. This traditional methodology is not a satisfactory base
on which to build our understanding of the earth's biological diversity.
We need to move from the scientists' preoccupation with 'publication' and
emphasize the opportunities offered by new methods of 'information
transfer'.

Collections as databases

The potential of collections as databases is not always appreciated. Ammann (1986) elegantly described the contribution that documented reference collections of dried flowering plants could make to biological research, and his thesis is equally applicable to fungi and invertebrates. However, if collections are to be regarded as essential databases they should be created purposefully, and not be accumulations of randomly acquired material. Collecting always involves a strong element of serendipity, but if the purposes of a collection are identified in advance then achievement of these purposes will be more likely. An objective might be to acquire representation of all the taxa of a group from an area, and to demonstrate the differences between sexual and immature stages. More comprehensively, the objectives might be to record the distribution, hosts, seasonality, and any associated variation of the included taxa. Such a collection could be enhanced by including symptoms of damage, galls or leaf mines produced, and could be extended to include karyotype, chromatographic profiles, electrophoretic banding pattern preparations, as well as voucher specimens from various experiments.

A data-rich collection of this sort, deliberately developed and carefully cross-indexed, is potentially a much more powerful tool than the haphazard accumulation of specimens found in most museums. Further, much of the contained data lends itself to computerization to facilitate access and transnational exchanges of information.

The value of dried reference collections for the elucidation of phylogenetic relationships has entered a new phase through the advent of molecular biology, especially the use of the polymerase chain reaction (PCR) to amplify DNA sequences. This has been achieved from herbarium reference specimens of fungi by Bruns *et al.* (1990).

In the case of microorganisms, a vital extension of the dried reference collections is one of permanently preserved living material, preserved by freeze-drying or in liquid nitrogen (Smith, 1988). A considerable number of filamentous fungi, and most bacteria and yeasts, cannot be named reliably unless living cultures are available, because the diagnostic features cannot be demonstrated in dead dried material. Culture collections are also the seed banks and botanical gardens of the microbiologist, able to supply living material when required. Live material is essential for studying disease response, biochemical systematics, verifying identifications, and for screening for beneficial properties. When deposited as vouchers, they also enable later workers to repeat or extend methods using the same strains.

The maintenance of living collections is costly both in labour and equipment, and it is clearly unrealistic to expect all species to be preserved in this way (the Noah's Ark principle). At the same time, it is salutary to reflect that while about 254 000 strains of fungi are preserved in the world's culture collections, these represent only 11 500 species; that is, 17% of the known and 0.8% of the estimated world species (Hawksworth, 1991a).

A crucial step in verifying identifications will remain the ability to match unnamed material to specimens or cultures already accurately named. This principle was recognized long ago by the International Mycological Institute's first mycologist (Mason, 1940). Major institutions with collections can assist those being developed elsewhere by exchanging, donating, or otherwise making available duplicates of authoritatively named specimens. This activity can be of major benefit to any institution starting to establish its own capability in identification.

Collections also provide the basis for much of our reference and information systems. Taxonomists, the people who work with collections, also work with published information. They have the experience to rationalize the wide range of publications, in many styles and languages, which bear on the diversity problem. Check-lists of species, whether systematic and comprehensive, or focused on host plants, on crops or on particular areas, form the basis for our comprehension of biodiversity. Despite this, and despite the importance of fungi and insects, there are no world lists of the described and accepted species apart from a few smaller groups. The Natural History Museum in London has card catalogues for the species names in most groups of insects, but to convert these into an electronic database and make them available to the rest of the world would require a major financial input. We now have the technical ability with computers to produce inventories of the world's organisms, together with cross-referencing to countries and host plants, to habitats and seasons. The botanists are leading the way in this exciting venture, most notably through the proposal to produce a *Species Plantarum* (Prance, 1992); a parallel *Species Fungorum* has been initiated at the International Mycological Institute. Linked with these initiatives are plans to accord the names accepted in such lists a specially protected nomenclatural status (Hawksworth, 1991b).

Zoologists and entomologists, with a vastly larger problem, are now actively considering how to tackle this same objective (Savage, 1990). However, they have yet to develop the essential international collaboration, attract the necessary resources, and above all to create the sense of purpose to meet this challenge of documenting biodiversity.

Collections and training

Collections also provide an important resource for training other biologists. Experienced taxonomists have tended to specialize on the production of monographic works; publications that summarize in considerable detail the available information about the characteristics and distribution of a particular group of organisms and include keys so that they can be identified. Such carefully researched publications are essential for the future of science; they provide the bedrock on which the next generation of specialists can stand.

However, the training of other scientists demands a wider range of outputs. Although the need remains for comprehensive introductions to particular groups, increasingly important are user-friendly keys to help other workers identify the organisms of their regions or of a particular crop. In addition to the traditional printed formats, expert-systems able to run on personal computers are opening new horizons. These different levels of publications and other outputs require different styles of writing, illustration, and presentation. The staff of the CAB International Institutes and those of the science departments at The Natural History Museum have now developed the expertise to organize training courses in biosystematics at various levels, as required by potential customers, from non-graduate technicians to postdoctoral research workers.

Most potent of all for the future, in the training area, are the relatively simple but accurate books, with excellent coloured illustrations and photographs, aimed at children and the general public. These, together with training videos and a variety of television broadcasts aimed at a larger market, will determine progressively the level of interest from the electorate and taxpayer concerning how much pressure to bring on governments for the protection of the environment.

Collections and the future

If the biodiversity of this planet is to have a future, then it is essential that we significantly enhance our knowledge, in breadth as well as in depth, of the organisms around us. This will not be possible, and certainly not be effective, if it is concentrated on a few Western centres. The number of useful data-rich collections within tropical countries needs to be increased, together with the number of suitably trained, indigenous, scientific staff. It is with this in mind that The Natural History Museum has recently signed an agreement with the new Biodiversity Institute of Costa Rica, to help train their staff, technicians and research workers, in a range of methods for collecting and studying insects and maintaining collections (*see p. 64*). Similarly, the CAB International Institutes organize training courses, in this country and overseas, to transfer expertise in biodiversity to other countries. Indeed the strengthening of national and regional capabilities in identification and taxonomy was accepted as a major focus for CAB International at its 1990 Review Conference.

In this way we aim to promote a two-tier scenario, whereby the major centres provide training and back-up identifications, collaborate with other countries in the production of reference works, identification guides, and even popular books on natural history, and advise on the most cost-effective ways of dealing with the vast number of undescribed species. No one museum or institute can possibly house representatives of all the world's organisms. No one centre is likely to be able to afford the books that would be required to house descriptions of all the world's organisms. The Entomology Library at the Natural History Museum currently comprises 85 000 volumes. If there are 10 million species of insects in the world and each requires two pages of descriptions, this would occupy 80 000 volumes (each of 250 pages and 2 cm wide) and 1.6 km of shelving, thus almost doubling this library's size just for this basic descriptive material. Similarly, if all the world's fungi were represented in the collections of the International Mycological Institute to the level they are at present that would require 70 km of shelving (Hawksworth, 1991*a*).

Collections and their associated libraries have a crucial role in our efforts to respond to the biodiversity crisis. Through these collections we can record and monitor the changes in the organisms around us. With them we can produce many products which are important in the sustainability of agricultural systems, the identification manuals, maps, and check-lists of pests. And from them we can derive the inspiration to produce new ideas on the origin, the significance and the maintenance of biodiversity. Particularly disturbing is the lack of awareness of the importance of collections even among some biologists (Clifford *et al.*, 1990) – although this has occasioned publicity in their defence (Anon., 1990).

But to be effective in communicating information, our investment in collections needs to be spread more evenly through various countries. Equally, to be effective our means of information transfer must become less dependent on the printed word. The next few years must see the application of electronic data handling on a much broader scale than we currently imagine. Computers are now more widely available than taxonomic libraries; electronic communication is even now cheaper than traditional methods of printer's ink on paper. If we accept the humanistic goal of a complete inventory of the world's biota, and most of us subscribe to this goal while realizing the logistic problems, then we will have to train a wider range of people in more countries, develop new collections with specific objectives, and adopt new methods of information transfer. Currently the CAB International Institutes and The Natural History Museum are collaborating in various ways, with each other and with institutes and universities in other parts of the world, in attempts to solve the problems which stem from the range of biodiversity on our planet. These are global problems requiring international co-ordination and effort on a level hitherto not contemplated.

References

Ammann, K. (1986) Die Bedeutung der Herbarien als Arbeitsinstrument der botanische Taxonomie. Zur Stellung der organismischen Biologie heute. *Botanica Helvetica* 96, 109–132.

Anon. (1990) In defence of taxonomy. *Nature* 347, 222–224.

Bruns, T.D., Fogel, R. and Taylor, J.W. (1990) Amplification and sequencing of DNA from fungal herbarium specimens. *Mycologia* 82, 175–184.

Clifford, H.T., Rogers, R.W. and Dettman, M.E. (1990) Where now for taxonomy? *Nature* 346, 602.

Coddington, J.A., Larcher, S.F., and Cokendolpher, J.C. (1990) The systematic status of Arachnida, exclusive of Acari, in North America north of Mexico. In: Kosztarab, M. and Schaeffer, C.W. (eds), *Systematics of the North American Insects and Arachnids: Status and Needs*. Virginia Polytechnic Institute and State University, Blacksburg, pp. 5–20.

Dayton, L. (1991) On the saving of species. *New Scientist* 129(1752), 25–26.

di Castri, F. and Younès, T. [eds] (1990) Ecosystem function of biological diversity. *Biology International, Special Issue* 22, 1–20.

Gaston, K.J. (1991) The magnitude of global insect species richness. *Conservation Biology*, in press.

Hawksworth, D.L. (1991*a*) The fungal dimension of biodiversity: magnitude, significance, and conservation. *Mycological Research* 95, 641–655.

Hawksworth, D.L. [ed.] (1991*b*) *Improving the Stability of Names: Needs and Options*. [Regnum Vegetabile No. 123.] Koeltz Scientific Books, Königstein.

Hawksworth, D.L. (1992) The need for a more effective biological nomenclature for the twenty-first century. *Botanical Journal of the Linnean Society*, in press.

Hodkinson, M.G. and Casson, D. (1991) A lesser predilection for bugs: Hemiptera (Insecta) diversity in tropical rain forests. *Biological Journal of the Linnean Society* 43, 101–109.

Hudson, H.J. (1986) *Fungal Biology*. Edward Arnold, London.

Mason, E.W. (1940) On specimens, species and names. *Transactions of the British Mycological Society* 24, 115–125.

Morris, M.G., Collins, N.M., Vane-Wright, R.I. and Waage, J.K. (1991) The effects of management on the invertebrate community of calcareous grassland. In: Collins, N.M. and Thomas, J.A. (eds), *The Conservation of Insects and their Habitats*. Academic Press, London, pp. 319–347.

Norris, K.R. (1974) General biology. In: *The Insects of Australia. Supplement*. CSIRO and Melbourne University Press, Melbourne, pp. 21–24.

Plotkin, M.J. (1988) The outlook for new agricultural and industrial products from the tropics. In: E.O. Wilson (ed.), *Biodiversity*. National Academy Press, Washington, DC, pp. 106–116.

Poinar, G.O. (1983) *The Natural History of Nematodes*. Prentice Hall, Englewood Cliffs, NJ.

Prance, G.T. (1992) The Species Plantarum Project. *Botanical Journal of the Linnean Society*, in press.

Price, P.W. (1988) An overview of organismal interactions in ecosystems in evolutionary and ecological time. *Agriculture, Ecosystems and Environment* 24, 369–377.

Quicke, D. (1991) Origin of a sponsored species. *New Scientist* 129(1753), 68.

Ride, W.D.L. (1991) Justice for the living: a review of bacteriological and zoological initiatives in nomenclature. In: Hawksworth, D.L. (ed.), *Improving the Stability of Names: Needs and Options.* [Regnum Vegetabile No. 123.] Koeltz Scientific Books, Königstein, pp. 105–122.

Savage, J.M. (1990) Meetings of the International Commission on Zoological Nomenclature. *Systematic Zoology* 39, 424–425.

Smith, D. (1988) Culture and preservation. In: Hawksworth, D.L. and Kirsop, B.E. (eds), *Living Resources for Biotechnology: Filamentous Fungi.* Cambridge University Press, Cambridge, pp. 75–99.

Stork, N.E. (1988) Insect diversity: facts, fiction and speculation. *Biological Journal of the Linnean Society* 35, 321–337.

Thatcher, M. (1989) *The Global Environment.* [Text of speech made by Prime Minister, The Right Hon. Margaret Thatcher M.P., to the United Nations General Assembly, New York, November 1989.] Prime Minister's Office, London.

Wilson, E.O. (1988) The current state of biological diversity. In: Wilson, E.O. (ed.), *Biodiversity.* National Academy Press, Washington, DC, pp. 3–18.

Wolf, E.C. (1987) *On the Brink of Extinction: Conserving the Diversity of Life.* Worldwatch Institute, Washington, DC.

Discussion

Bazin: The positive correlation between diversity and ecosystem stability seems to have been assumed, and I wondered if you had any actual evidence that this was indeed the case?

Hawksworth: In the case of microorganisms and invertebrates it depends on what communities are being considered. In leaf-litter, for example, many species may have similar roles and so functional redundancy occurs and in some instances perhaps several hundred species might be lost with little effect. However, if one was considering a species of fungus which was the sole mycorrhizal symbiont of a dominant tree or a single pollinating insect of that tree, the loss of even single 'keystone' species could result in the collapse of a whole community. Further research is needed before generalizations can be made.

Bazin: On a theoretical basis, at least for linearly connected sytems, the probability of stability decreases as the number of species, and the interactions between them, increases. Thus, *a priori*, there is no support for the assumption. What is required to test this are measurable estimates of stability and diversity and appropriate experimentation. Observations alone are unlikely to resolve such fundamental questions.

Bull: In discussing the numerical values of biodiversity some consideration needs to be given to the emphasis which is accorded to the term 'species', particularly as applied to microorganisms. The inventory of microbial diversity is likely to be very incomplete, and the often quoted figure of 4760 species of bacteria, cyanobacteria and mycoplasmas [Wilson, E.O.

(ed.), 1988, *Biodiversity*. National Academy Press, Washington, DC.] an enormous underestimate. The species concept for prokaryotes is much less definitive than for larger organisms and it seems preferable to discuss biodiversity in terms of genetic diversity. The definitions of biological diversity adopted by McNeely *et al.* (McNeely, J.A., Miller, K.R., Reid, W.V., Mittermeier, R.A. and Werner, T.B., 1990, *Conserving the World's Biological Diversity*. International Union for the Conservation of Nature, Gland) and the US Congress Office of Technology Assessment (*Technologies to Maintain Biological Diversity*. US Government Printing Office, Washington, DC, 1987) encompass ecosystems, species, and genes. Such a definition highlights the genetic variation which is evident in microbial species, and also the necessity of preserving whole ecosystems in order to maximize the conservation of microbial gene pools.

Four

Problems of Assessment of Biodiversity

J. Boussienguet, *National Biological Control Programme, Université du Gabon, P.O. Box 1886, Libreville, Gabon*

ABSTRACT: While significant progress has been made in the field of molecular biology and bioengineering, this development is taking place in a way as if it was possible to manipulate genes without the species – the raw material for biotechnology. Biosystematics have declined because we focus on the ethical and emotional aspects of biodiversity, whereas we should emphasize the contribution of taxonomy in successful applications to agriculture, medicine and industry, in goods production and resource conservation. Concern is expressed at the gap between supply and demand for taxonomists in developing countries, the main centres of biodiversity. Attention is drawn to the erosion of genetic resources, to the magnitude of biodiversity and to revised estimates of living species number: with only 5% of the living species described so far, the task ahead is enormous and we should establish internationally agreed priorities, develop identification services and training courses in taxonomy and avoid duplication of the resources offered.

The 'civilization of the gene', to which we are committed by the end of this century, offers unprecedented prospects for agriculture and industry. However, this promising development in the field of bioengineering is of a precarious nature. Indeed, whereas significant progress has been made in the field of manipulating the infinitely small, the gene, the erosion of genetic resources is of increasing concern. Our knowledge of biodiversity, the raw material for biotechnologies and for agricultural development, is still in its infancy.

The leading position occupied by molecular biology during recent decades, and, more recently, the coming into fashion of ecology, have relegated

The Biodiversity of Microorganisms and Invertebrates: Its Role in Sustainable Agriculture.
Edited by D.L. Hawksworth. © CAB International 1991.

botany and zoology to the rank of impoverished parents among the scientific disciplines. In many countries which represent the cradles of biosystematics, support for taxonomy is declining at an alarming rate in universities and from international and national grant-awarding bodies and research councils. The means at our disposal for increasing sustainable agricultural production multiply rapidly thanks to genetic engineering and biotechnology, but these developments are taking place as if it were possible to manipulate genes without the species, or to influence ecology while ignoring animals and plants.

A lack of funding is not the only obstacle to the assessment of biodiversity. The present decline of biosystematics may also be due to our own failure to properly answer the question: why are the classification, description and naming of variations among organisms useful from a practical point of view? Useful not as an esoteric exercise for 'taxonomist lawyers', but as a tool to be utilized for modern agriculture, medicine, industry, and energy production.

Agronomists, molecular biologists, genetic engineers, etc., require taxonomic knowledge, and they require that for as many organisms as possible, and as soon as possible. Biosystematics has declined because the contribution of taxonomy to successful applications of biological control, gene transfers, etc., to the production of goods and the protection of resources have not been adequately stressed. This failure to demonstrate the commercial value of taxonomy to donors and decision-makers, in the quantification of the economic value of biosystematics services, is of especial concern. In fact, the commercial value of biosystematics services is equal to, if not higher than, the market-place values represented by bioengineered goods. But how can the public awareness of the biosystematics impact be promoted?

It is pertinent to mention here, the peculiar nature of the taxonomist's profession. Indeed, even in countries with an ancient scientific tradition, there are no recognized first degree university courses, and very few at the masters level, on completion of which a degree is conferred which leads to the profession of taxonomist. In most countries, including the French speaking ones, biosystematics has lost credibility in the academic world. Only a small number of professionals contribute to it, and the survival of the discipline depends on the passions of individual scientists and the personal carvings made by each taxonomist into the living world. Consequently, while some classes or organisms are well-studied, others are completely abandoned for the lack of taxonomists, a threatened species themselves.

The gap between the supply of and demand for taxonomists is widest in developing countries. This is especially unfortunate as tropical countries are the main centres of the world's biological diversity. For example, in an area of about 15 hectares of rainforest in Borneo, about 700 species of trees were identified, as many as in all of North America. While covering only 6% of the earth's land surface, tropical forests contain at least 50% of all the earth's species. But how many areas of the world, even in Europe and North

America, have up-to-date basic faunas, floras, or even check-lists for all groups? There is no hope of remedying this situation without a genuine effort to accord recognition to the value of taxonomy, and to develop identification services and training courses in taxonomy.

Another difficulty for the assessment of biodiversity is the countless number of living species. How many species are involved? According to the United Nations Environment Programme (UNEP, 1987), there are between 1.7 and 30 million living species, the great majority in tropical forests, and a substantially larger number of invertebrates than vertebrates. Needless to say, new birds and mammals are being continually discovered, at a rate of perhaps one mammal and three birds per year, mainly in tropical regions. Although fairly reliable inventories are available for mammals, the biodiversity of invertebrates and microorganisms which still remains to be described is a challenge to be overcome. The proportion of species described is far below a desirable number (see Hawksworth and Mound, Chapter 3).

Assuming that there are three times as many species in the tropical as in the temperate zones, as in birds and mammals, the total number of living species could be estimated at 5–6 million, according to the 1.7 million already described, mainly from temperate zones and including 25 000 protozoans, 20 000 worms, 100 000 molluscs and 925 000 arthropods. A dramatic upward revision of this estimate, which was regarded as adequate until the end of the 1970s, has resulted from Erwin's (1983) careful studies of the arthropod fauna in the canopy of tropical trees in Panama. Using an insecticidal fog to knock down the canopy arthropods, Erwin showed that the major part of this fauna is confined to the tree canopy. This is hardly surprising because it is the summit of the tree that receives the maximum sunlight and carries most of the flowers, fruits, and green leaves. From the revolutionary work of Erwin, it has been established that the canopy of tropical trees carries three times as many species as does the undergrowth, and that the fauna specific to each tree species consists of approximately 600 arthropod species. Assuming there are about 50 000 species of tropical trees, the total number of tropical arthropod species would amount to 30 million and as arthropods represent the major part of all living species, it can be concluded that earth's biodiversity consists of over 30 million species.

The different arguments on which these calculations are based can be debated, especially the figure of 600 for the number of arthropods specific to each tropical tree species. It is more important to design experiments which would enable these estimates to be confirmed, or otherwise. Biologists have realized the importance of these studies to the extent that currently there are several teams competing to build the most ingenious device for gaining access to study the fauna of the tropical tree canopy.

Of a conservatively estimated 30 million living organisms, we have so far described only about 5% at a rate of 15 000–20 000 species per year. At that rate, we will need almost 2 more millenia even to approach the description of a reasonable percentage of living organisms. Do we have so much time before many species become extinct? Unfortunately, in many tropical centres of biodiversity, the erosion of genetic resources as a result of the destruction of tropical forests continues at a greater rate than studies of biodiversity. Different assessments of the ratio of deforestation have been reported in the last few years. On a global basis, the world's forests are disappearing at the rate of 15 million ha year^{-1} (UNEP, 1987), with most losses occuring in the humid parts of Africa and Latin America. At the present rate of deforestation, about 40% of the remaining forest cover in developing countries will be lost by the year 2000 (Organization for Economic Co-operation and Development, 1985). For example, in the Ivory Coast, more than 70% of the area under forest has been destroyed in the last 70 years. According to Myers (1988), in Nigeria the moist forest cover has been reduced by at least 90%.

If we consider medicinal plants, it has been calculated that 5% of all tropical forest plant species serve the cause of medicine (Jacobs, 1982). Given the present and projected rates of forest destruction, it can be estimated that several thousand species are being eliminated each year (Ehrlich and Ehrlich, 1981). According to an estimate by Wilson (1985), if a forest area is reduced to 10% of its original size, the number of species that can continue to exist in it will eventually decline to 50%. For example, monitoring of biodiversity as an indicator of forest change has shown that 96 of the 380 European species of butterflies are threatened with extinction (Pavan, 1986).

In this race against time, there is a need to establish internationally agreed priorities, help developing countries to develop their own taxonomic research centres, and avoid duplication of effort and resource utilization.

In some developing countries, significant progress is being made in reafforestation. But as far as biodiversity is concerned, timber plantations do not mean forests. In theory, there exists a will among nations to preserve the natural environment and to save the maximum possible number of species from extinction. However, with few exceptions, this policy has had little impact. As we know, reserves are often located in areas chosen mainly for convenience or expediency rather than for their wealth of species which, it must be emphasized, is largely unknown. Indeed, if we do not have adequate species inventories for the protected areas, what are we aiming to conserve, and how can we be sure we are conserving it?

Finally, in addition to obstacles related to the large diversity of living species, the lack of funding and human resources, and the erosion of genetic resources, there are also obstacles inherent in methodology. For example, differences in approach between the assessment of biodiversity in terrestrial as opposed that of marine environments. From a practical point of view, it must be emphasized that in order to rally more support for the assessment of biodiversity, this research has to be of urgently needed practical value. There is competition with genetic engineering, molecular biology, biological control, etc., for funds, status, and credibility. Donors and decision makers must be convinced by something other than our almost *quasi*-religious approach on the ethical and emotional aspects of biodiversity.

References

Erwin, T.L. (1983) Biodiversity of tropical tree canopy. *Bulletin of the Entomological Society of America* 2, 14.

Ehrlich, P.R. and Ehrlich, A.H. (1981) *Extinction: The Causes and Consequences of the Disappearance of Species*. Random House, New York.

Jacobs, M. (1982) The study of minor forest products. *Flora Malesiana Bulletin* 35, 3768–3782.

Myers, N. (1988) Tropical forests: much more than stocks of wood. *Journal of Tropical Ecology* 4, 209–221.

Organization for Economic Co-operation and Development (1985) *The State of the Environment 1985*. Organization for Economic Co-operation and Development, Paris.

Pavan, M. (1986) *Charte sur les Invertébrés*. Conseil de l'Europe, Comité des Ministres, Brussels.

United Nations Environment Programme (1987) *The State of the World Environment*. [Report No. L14/16.] United Nations Environmental Programme, Nairobi.

Wilson, E.O. (1985) The biological diversity crisis: a challenge to science. *Issues in Science and Technology* 2, 20–29.

Discussion

F. Baker: Assuming that there are about 20 million insects not yet described, as suggested could be the figure by Hawksworth and Mound (Chapter 3), I wonder how long would it take to describe all 20 million and how that figure compares with the rate at which undescribed insect species are disappearing.

Boussienguet: A taxonomist can describe about 5 000 insects in the course of a lifetime, and therefore 20 million divided by 5 000 provides an indication of the magnitude of the task, that is around 4 000 lifetimes.

F. Baker: The problem of the numbers of taxonomists and their training has to be taken into account in relation to the rates of disappearance of the species. These questions are often discussed but never resolved.

Five

The Dimensions of Biodiversity: The Use of Invertebrates as Indicators of Human Impact

J.D. Holloway, *International Institute of Entomology, 56 Queen's Gate, London SW7 5JR, UK*
N.E. Stork, *Department of Entomology, The Natural History Museum, Cromwell Road, London SW7 5BD, UK*

ABSTRACT: The profile of insect biodiversity is reviewed and its disruption through human utilization and management assessed. Absolute diversity, the numbers of species present, is discussed with relevance to the various methods of estimation adopted. The generic, biogeographic, and ecological dimensions of biodiversity are distinguished, including the relevance of ecosystem structure, body size, and trophic complexity. Particular attention is paid to the use of invertebrates as environmental indicators, drawing mainly on experience with moths in South-East Asia. The priority needs for maintaining biodiversity through sustainable agriculture are seen as: clarification of the dimensions of biodiversity to obtain a clearer picture of its natural profile, assessment of the changes that occur when we modify or attempt to change that profile, adoption of a standard set of sampling methods to facilitate comparisons between different parts of the world, and the use of tried and tested indicator groups. The full involvement of a critical mass of specialists with international broad-based reference collections is required. Finally, agronomists are encouraged to take note of the value of biodiversity when advising on the economic advantages of different schemes.

Introduction

In the last 5 years a new term, 'biodiversity', has come into common usage, encompassing not only the immense richness of life on this planet but our concerns over its destruction. This richness of life is not uniform but is heterogeneous with respect to a number of parameters. These are often

The Biodiversity of Microorganisms and Invertebrates: Its Role in Sustainable Agriculture.
Edited by D.L. Hawksworth. © CAB International 1991.

restricted to three: genetic diversity, species diversity, and ecosystem diversity (e.g. International Union for Conservation of Nature *et al.*, 1990). We suggest that the picture is more complex and multidimensional, including geographic, temporal (successional and long-term changes), structural, communal (trophic networks), and fractal (diversity of scale) components.

Invertebrates, and in particular insects, comprise the vast majority of life-forms on earth, and our paper will concentrate on this group. Swift and Anderson (1989) divided ecosystems into three major components: the plant subsystem, the grazing subsystem, and the decomposer subsystem. Invertebrates predominate in the second and third, and also in a fourth, the predator subsystem.

We review here what is known of the natural profile of biodiversity and assess the disruption and modification of that profile through our utilization and management of natural resources. Because of the immense diversity of life in most biotopes, it is often necessary to study 'indicators' of change rather than the whole biota and we shall review the importance and usefulness of insects in that role. We restrict our arguments to terrestrial systems, although they are applicable in most senses to freshwater and marine systems. In these systems nematodes and polychaetes have been successfully used as indicators of environmental change, such as disturbance and pollution (e.g. Lambshead *et al.*, 1983; Grassle, 1989).

Our approaches to the study of biodiversity have been from opposite ends of the scale, N.E.S. focusing on global diversity and the structure of whole ecosystems (e.g. Stork, 1988; Stork and Gaston, 1990), and J.D.H. using a single group, the butterflies and moths, as an indicator of biogeographic and ecological pattern in diversity.

Absolute diversity

Most elements of the periodic table have been known for some time and this knowledge has been essential in advancing the fields of chemistry and molecular physics. But it is surprising that in an age when vast sums of money are being poured into determining how many stars there are in the solar system, we devote such little effort to answering a similar question far closer to home: how many species of living organisms are there on earth? The number of stars remains more or less static in human perception, but the number of species may be diminishing as rapidly as we try to determine it. Each year less than 10 000 species of organisms are described as new and perhaps 3 000 existing species are placed in synonymy, yet recent predictions (based on a world total of 5–10 million) suggest that species extinctions may approach 17 500–35 000 a year (Wilson, 1988).

The approach by most early biologists to this question was to describe species as rapidly as possible. For some groups such as birds this task is

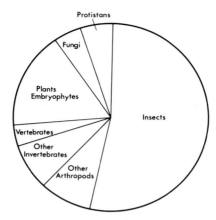

Figure 5.1. Proportions of major groups of organisms based on a total of 1 412 000 described living species (after Barnes, 1989; with figures for microorganisms from Di Castri and Younès, 1990).

almost complete with only one or two species being added each year to the total of about 9 000. Estimates of the total number of named and described species of all organisms range from about 1.4 (Wolf, 1987; Barnes, 1989) to 1.8 million (Stork, 1988; Fig. 5.1). Discrepancies arise because past taxonomic efforts prove to have been inadequate. Some 'species' prove to be complexes of many species, and others, perhaps highly variable morphologically, have been described many times. Thus, taxonomists with a good grasp of the world fauna of their specialist group and with access to a world collection, often may publish more synonymies than new descriptions!

Current interest in absolute numbers of species was stimulated by Erwin (1982) whose 'back of the envelope' calculation pointed to a total of 30 million tropical forest arthropod species – an order of magnitude greater than the previous best estimate. Erwin, an eminent beetle taxonomist, used an extrapolative method based on ecological criteria. By fogging the canopies of 19 *Luehea seemannii* trees in Panama with insecticide, he collected and then sorted over 1 200 species of beetles. Using a range of guesses for host-specificity for different trophic groups, he estimated that an average 13.5% of these – 163 species – are host-specific to *L. seemannii*. To arrive at the total of 30 million he then assumed that all 50 000 tropical tree species had the same number of host specifics, that beetles comprise 40% of the canopy species, and that there are twice as many species in the canopy as on the ground. In some senses the assumptions Erwin made are as important as the global figure, and a number of authors have examined and refined them (Stork, 1988; Hammond, 1990; Stork and Gaston, 1990; Thomas, 1990).

An alternative, more empirical approach is to try and extrapolate from the diversity profile of a well-known group, such as the birds (Diamond,

1985) or butterflies. For example, of 22 000 insect species recorded in Britain 65 are butterflies. There are estimated to be a maximum of 20 000 butterfly species world-wide. Therefore, assuming British proportions to hold world-wide, there are 6.6 million species of insects in the world. Such an approach has yet to be tested on a broader regional or taxonomic scale. How consistent are the proportions of the major groups in faunas from different parts of the world? How similar are the distribution patterns of, say, the butterflies to those of other groups of insects? Numerous biogeographical reviews show that the distribution of insect diversity with altitude, latitude, climate and interactions with vegetation structure, floristics and chemistry show both commonalitites of pattern and variability (Gauld, 1986; Holloway, 1987).

A third and largely untested approach to estimating global diversity has a strong basis in theoretical ecology (May, 1988). Some biologists are studying patterns in food webs, host–parasitoid and predator–prey relationships, and the distribution of body size among organisms to enable further extra-polations to be made. In practice, they largely have addressed problems of community structure and have yet to turn these ideas towards providing global estimates.

To summarize, our current best estimate of global diversity at the species level through extrapolative methods is in the range of 5–10 million species, though Hodkinson and Casson (1991), extrapolating from Sulawesi insect data, obtained a figure of only 1.84–2.57 million insect species.

The genetic dimension

In attempting to assess absolute diversity through the study of pattern within it, we conventionally refer to species. But this species diversity has a genetic basis. Assessment of diversity at the level of species implicitly recognizes a convenient level of heterogeneity in the distribution of genetic diversity borne by individual organisms. Our appreciation of this heterogeneity is based squarely on the science of systematics, though systematists themselves frequently challenge the view that the taxonomic rank 'species' has any unique qualities to distinguish it from any other taxonomic rank (Nelson, 1989). Those that are familiar with such organisms as viruses, bacteria, and organisms that primarily reproduce asexually through parthenogenesis or other means, will have some sympathy for this alternative view.

Preston (1962) observed that individuals are distributed among species in a log-normal manner, and that species are distributed similarly among genera, and genera among families. It is possible that such a relationship exists between any pair of taxonomic ranks, in a sort of fractal taxonomic sequence, and that the distribution of genetic diversity could also be encompassed in the model, a concept that is addressed to some extent by

Perfect (Chapter 12). It is still a moot point whether the log-normal distribution has some biological significance or is merely a consequence of the statistical central limit theorem applied to the multiplicative effects of the many largely independent factors that control species abundances (Holloway, 1977; Sugihara, 1980; May, 1981).

The biogeographic dimension

The heterogeneity of the surface of the earth and its interaction with latitudinal effects and meteorology has provided a complex template on which biological diversity has developed through time. The resulting patterns are a synthesis of space, time and form as expressed by Croizat (1964). Their analysis and interpretation are similarly complex, and are dependent on sound systematics.

Area factors are important in determining species diversity (Preston, 1962; MacArthur and Wilson, 1967). Species numbers tend to double with a 10-fold increase in area, be it in relation to actual areas such as Pacific Islands (e.g. Robinson, 1975) or some similar concept such as plant–host range (see Strong *et al.*, 1984). From a knowledge of such area relationships it is possible to make predictions of extinctions when areas of habitats are reduced or the range of a plant is curtailed (e.g. Diamond and May, 1981), though the concept is not without its critics (e.g. Gilbert, 1980).

Species diversity increases dramatically with reducing latitude, but is not uniform across all groups of organisms (Gauld, 1986; Holloway, 1987). Variance from group to group in response to latitudinal gradients and to geographic isolation in conjunction with the area effect can be illustrated by the moth families in the Indo-Australian tropics (Holloway, 1987). Diversity of families increases with species richness as one moves from temperate regions, here represented by Japan, to the tropics, represented by Borneo, but then declines as one passes eastwards to the smaller and more isolated archipelagos of the Pacific (Fig. 5.2).

Diversity is strongly patterned on a geographic scale within each group, reflecting the past biogeographic and climatic history of a region. For example, the Indian butterfly fauna is drawn from several geographic types of genera, each with a different focus of species richness or massing centre (Holloway, 1969, 1974). The major component is of species from oriental tropical genera. These have a centre of species richness extending from the north-east Himalaya to Sumatra and Borneo. There are also contributions from genera with centres of richness in the savannas of Africa, in the Mediterranean region, and in western China and the western Himalayan zone of the Chitral and Pamirs. The species from these generic types segregate within India to some extent on ecological grounds: tropical forest, savanna, semi-arid and montane (Holloway, 1974; see also Pearson and Ghorpade, 1989, for Cicindelidae).

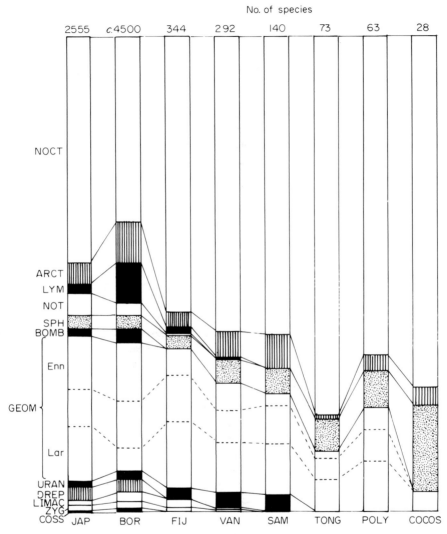

No. of species

Figure 5.2. Estimates of numbers of species (figures at tops of columns) and proportions of macromoth families in the faunas of JAPan, BORneo, FIJI, VANuatu, SAMoa, TONGa, a typical high volcanic POLYnesian island and COCOS-Keeling Atoll (Indian Ocean). The moth families are: NOCTuidae, ARCTiidae, LYMantriidae, NOTodontidae, SPHingidae, BOMBycoidea (superfamily), GEOMetridae (with subfamilies Ennominae and Larentiinae indicated), URANiidae (including Epiplemidae), DREPanidae, LIMACodidae, ZYGaenidae, COSSidae (after Holloway, 1987).

The uniqueness of certain floras and faunas depends on such biogeo-graphical patterns and historical factors. Britain, for example although isolated from the rest of Europe by the North Sea, the Bay of Biscay, and the Channel, has almost no endemic species. In contrast, New Zealand, at similar latitudes, and also separated by sea from Tasmania and Australia, has a large proportion of endemics. Extinction of the entire British fauna would result in little loss to global diversity provided similar habitats to those in Britain remain in other parts of Europe. Loss of the New Zealand fauna and flora, in contrast, would be devastating. It is important to identify such 'hot spots' of endemism (Myers, 1988, 1991) and ensure that they reflect real hot spots unbiased by varying levels of collecting in different regions (e.g. International Union for Conservation of Nature, 1987). Conservationists, be they local, regional, governmental or intergovernmental, are faced with hard decisions about what to save and about what not to save. Where only data on species distributions are available, statistical methods are available that will enable key areas to be identified (Holloway, 1984a). New methods are being devised that will allow the use of systematic information, where available, to provide a more sophisticated evaluation of critical areas or ecosystems (Ackery and Vane-Wright, 1984; Vane-Wright *et al.*, 1991; Collins and Morris, 1985). This approach is an essential first step in conservation planning, and a number of invertebrate groups, particularly butterflies, bees and tiger beetles (Cicindelidae), have considerable potential in providing the required level of data.

What is the effect on the biogeographic pattern in biodiversity of conver-sion by humans of natural habitats to managed systems, such as plantation monoculture or field-crop agriculture? This question may be answered using examples from Lepidoptera. Many species of this group are associated with natural, usually forested, ecosystems in the tropics and are often highly localized geographically. In the Indo-Australian tropics, for example, endemism on single island or island groups is very common, and there are recognizable major discontinuities in species distributions such as the lines of Wallace and Weber (Holloway, 1973, 1984b; Holloway and Jardine, 1968; but see Vane-Wright, 1991). Field-crop pest species, on the other hand, are drawn from an association of disturbed-habitat specialists that are mobile (often long-distance migratory), opportunist, and geographically very wide-spread. Figure 5.3 shows the summed distributions of the widespread com-ponent of the macro-Lepidoptera faunas of islands of the south-west Pacific (Holloway, 1979). This component makes up at least one-third of the faunas of these relatively remote islands, a proportion that rises to 60–80% on smaller or more isolated islands such as the Tonga group or Norfolk Island, and even higher on atolls such as Cocos-Keeling (Holloway, 1987). Field-crop pests included are, for example, noctuids from the genera *Achaea*, *Agrotis*, *Anomis*, *Earias*, *Chrysodeixis*, *Condica*, *Helicoverpa*, *Ophiusa*, *Othreis*, and *Spodoptera*.

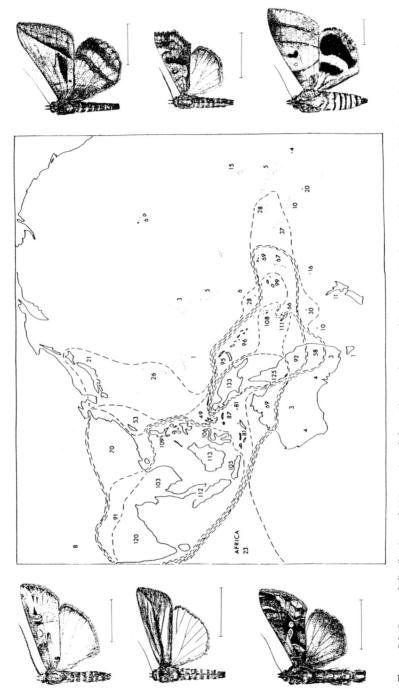

Figure 5.3. Summed distributions of 133 species of the widespread geographic element in the moth faunas of archipelagos of the south-west Pacific (from Holloway, 1979). Typical pest noctuid moths in this element are illustrated on each side: top to bottom, left, *Agrotis ipsilon*, *Mythimna loreyi*, *Chrysodeixis eriosoma*; top to bottom, right, *Mocis frugalis*, *Anomis flava*, *Ophiusa coronata*. Scale bars = 1 cm (moth portraits by B. Hargreaves).

It is evident from this that conversion to field-crop agriculture poses a threat to biogeographic diversity by reducing faunal heterogeneity. In arboreal cropping systems, such as plantations, this effect is not so serious as the pests are mostly drawn from forest specialist genera whose localization is more frequent: the limacodid moth pests of palms in South-East Asia show significant endemism (Cock *et al.*, 1987). However, conversion to plantation leads to loss of floristic and structural diversity in the vegetation to an extent similar to that of field-crop agriculture, and this is reflected in a loss of invertebrate diversity as discussed below.

The ecological dimension

The biogeographic patterns, or gamma-diversity (see also Lee, Chapter 7), are reflected to some extent on a local scale in beta-diversity, or between-habitat diversity patterns: over climatic or altitudinal gradients, and in zonation, biotope segregation or life-zones within a small area. Variation of family proportions in moth samples over an altitudinal transect within the tropics bears a close resemblance to that with latitude (Holloway, 1986, 1987). Most national resource management planning has to consider the pattern of beta-diversity within the country concerned, which has implications when considering invertebrates as indicators (p. 52).

Alpha-diversity, or within-habitat diversity, is concerned with diversity on a very local scale, though through consideration of ecological continua and succession, and overlapping associations of species, it is often difficult to isolate it from beta-diversity (e.g. Holloway, 1989), and phenomena such as the area effect discussed earlier. But there are a number of aspects of alpha-diversity that are largely independent of these factors such as the relationship of diversity to ecosystem structure, the distribution of body size among organisms, trophic relationships such as food webs and interactive guild ratios.

Such phenomena have been investigated by teams from The Natural History Museum in arthropod sampling programmes in the Indo-Australian tropics: north Borneo, north Sulawesi (Table 5.1), and Seram. The results indicate the immensity of the task in hand. For example, in insecticide 'fogging' samples from the canopies of 10 Bornean trees 1 455 individuals of chalcid wasps were obtained, representing 739 species, mostly undescribed. The commonest species was represented by 19 individuals and 400 by singletons. The complete arthropod fauna from this Bornean forest canopy shows the dominance of Diptera, Hymenoptera, and Coleoptera (Table 5.2), with ants dominant in terms of individuals and parasitic Hymenoptera in terms of species (Stork, 1991). Samples from elsewhere in the tropics show the same basic pattern (Erwin, 1983; Adis *et al.*, 1984; Adis and Schubart, 1985; Stork and Brendell, 1990).

J.D. Holloway & N.E. Stork

Table 5.1. Mean number of arthropods (M) and standard deviations (SD) collected in m^2 trays from natural forest and tree crops in lowland and upland areas in north Sulawesi by knockdown insecticide sampling.

| | Low altitude (approximately 200 m) | | | | | | Higher altitude (approximately 1100 m) | | | | | |
| | Forest | | | Agricultural | | | Forest | | | Coffee | | |
	M	SD	%	M	SD	%	M	SD	%	M	SD	%
Coleoptera	5.3	3.7	4.4	13.1	6.2	7.5	49.5	16.0	20.9	27.3	7.9	8.7
Diptera	92.6	21.6	70.3	59.8	47.6	28.7	70.5	16.5	30.7	37.8	17.8	12.2
Formicidae	3.2	5.7	2.2	78.7	118.0	26.4	11.0	29.2	4.1	219.0	376.0	32.4
Hymenoptera	10.5	8.0	9.4	11.5	5.7	6.7	20.2	4.5	8.6	9.5	4.5	3.6
Lepidoptera	4.4	3.2	4.2	2.7	2.3	1.3	6.5	2.5	2.8	10.5	6.4	3.1
Hemiptera	6.1	5.0	5.5	9.0	5.8	5.0	9.0	6.1	3.7	11.8	5.3	3.7
Orthoptera	1.3	1.8	1.0	4.7	6.4	2.2	8.0	5.8	3.2	11.7	5.1	3.5
Other insects	1.7	1.5	1.7	64.0	116.7	15.9	23.0	10.7	9.7	101.3	63.9	28.9
Other arthropods	1.7	1.6	1.2	15.2	13.6	6.2	38.0	12.2	16.4	13.6	6.9	4.0
Total Mean	126.8			258.7			235.7			442.5		
SD	32.6			237.6			43.1			365.0		

Table 5.2. Total numbers (*n*) of individuals and species of arthropods collected from 10 lowland forest trees in Borneo by knockdown insecticide fogging. SD/M = standard deviation divided by mean across 10 trees, M% = mean % across 10 trees. Note species totals are not available for Braconidae, Psocoptera and other arthropods therefore the final total for species is incomplete

	Individuals			Species		
	n	SD/M	M%	*n*	SD/M	M%
Diptera	5046	0.48	21.5	444	0.21	21.6
Formicidae	4429	0.80	18.2	98	0.34	3.1
Coleoptera	4000	0.51	16.9	859	0.43	21.4
Hemiptera	2628	0.65	11.2	268	0.28	7.5
Other Hymenoptera	2117	0.52	8.8	945	0.46	22.7
Arachnida	1138	0.45	4.9	193	0.35	6.2
Psocoptera	888	0.43	3.8	?	0.28	3.6
Thysanoptera	851	1.13	3.1	130	0.41	3.8
Collembola	736	0.60	3.2	22	0.41	1.7
Lepidoptera	617	0.48	3.2	50	0.23	2.8
Orthopteroids	542	0.36	2.5	86	0.30	3.0
Blattodea	440	0.72	1.9	40	0.55	2.1
Other insects	68	0.72	0.3	21	0.69	0.5
Other arthropods	85	0.90	0.5	?	?	?
Total	23482			3059		

Ecosystem structure

The canopy is but one of the (arguably) five major biotope components of tropical forests: soil, leaf litter, herb-layer, tree trunks, and canopy. In a study of the community structure of these biotypes in lowland forest in Seram, all five were sampled using a range of direct sampling methods (Stork, 1988; Stork and Brendell, 1991). Overall abundance, biomass and the proportions of the different arthropod groups vary widely between these biotopes (Figs 5.4, 5.5 and 5.6), as can be seen particularly for ants, mites, springtails, and 'other arthropods' (i.e. arthropods other than insects and arachnoids, mostly Diplopoda). The segregation of species between these biotopes (and other biotopes not considered in this experiment) is much more difficult to assess. Detailed sorting, largely by P.M. Hammond, of over a million individual beetles collected by a wide range of sampling methods on the ground and in trees in north Sulawesi, suggests that, from over 5 000 beetle species that resulted, considerably less than half are true canopy species (Hammond, 1990; P.M. Hammond, N.E. Stork and M.J.D. Brendell, personal observations). This is in sharp contrast to Erwin's (1982) suggestion, mentioned above, that two-thirds of forest arthropod species are in the canopy.

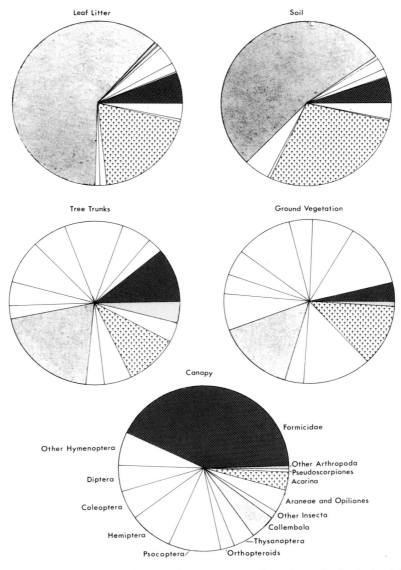

Figure 5.4. Pie-charts showing the contribution in terms of numbers of individuals of the major arthropod groups in samples taken from soil, leaf litter, ground layer vegetation, tree trunks and canopy in lowland rain forest in Seram (from Stork and Brendell, 1991).

Although segregation of diversity among these biotopes in terms of species is still under investigation, the impact of conversion to field-crop agriculture is easy to imagine. Diversity dependent on the arboreal components (including decomposition biotopes such as leaf litter and fallen trees)

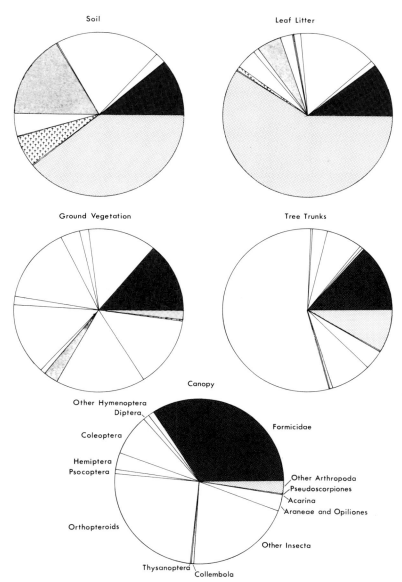

Figure 5.5. Pie-charts showing the contribution in terms of biomass of the major arthropod groups in samples taken from soil, leaf litter, ground layer vegetation, tree trunks and canopy in lowland rain forest in Seram (from Stork and Brendell, 1991).

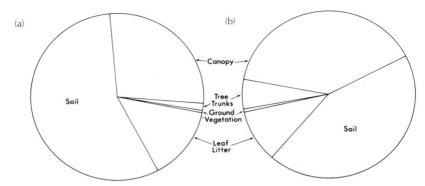

Figure 5.6. Pie-charts showing the relative contributions of the major groups of arthropods and arthropod faunas of the five main biotopes in a hectare of forest as extrapolated from data from sampling in Seram lowland rain forest: individuals (a) and biomass (b) of different biotopes (from Stork and Brendell, 1991).

will be eliminated. The under-storey or ground flora will be simplified, particularly with monoculture. The litter layer will be reduced or will disappear, and it and the soil component will be modified through exposure to greater climatic variability and through disturbance to nutrient cycles.

Comparable data from managed systems are very few, and effort should be made to obtain them. Preliminary data from north Sulawesi (Stork and Brendell, 1990) on the relative numbers of different groups of insects show differences in composition in the faunas of natural lowland forest and mixed arboreal crops such as coconut, and clove. Diptera are less abundant and ants and Collembola more abundant in the managed systems (Table 5.1). There is also a slight reduction in the numbers of parasitic Hymenoptera. Similar results were obtained at higher altitudes with coffee plantations. Examples from light-trap catches of Lepidoptera of changes within an insect group will be given later.

Body size

The mathematical relationship between body size and number of species is well-known (May, 1981, 1986). Progressively smaller body size classes contain progressively greater numbers of species, that is until both the limiting lower size of multicellular organisms and that of sexual reproduction breaks down. Whether this relationship continues in some form for bacteria, viruses and smaller organisms is uncertain. Similar patterns have been found and are being examined for natural insect communities on plants (Morse *et al.*, 1988; N.E. Stork, unpublished). Morse *et al.* (1985) found that the abundance of arthropods of different sizes was related to the fractal dimension of the plants they had been sampled from. In other words, plants with

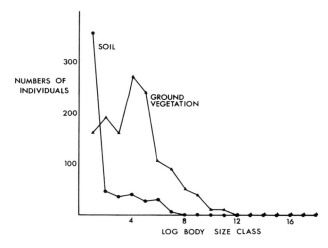

Figure 5.7. The abundance of arthropods with body size (on a logarithmic scale) in above- and below-ground samples from lowland forest in Seram, as exemplified by data for ground vegetation and soil respectively (from Stork and Brendell, 1991).

more patterned leaves and bark provide more niches, and a wider range of niches, for different insects to occupy.

The relationship of size distribution to the structural components of the ecosystem is illustrated from the Seram data in Fig. 5.7. Comparison of above- and below-ground invertebrate faunas, here exemplified by the size distributions of arthropods from samples from ground vegetation and soil, respectively, indicates a significantly lower peak for the soils and leaf litter faunas. One might expect this from the different fractal dimensions of the two classes of habitat: the interstices of the soil versus the lattices of the forest above. Microorganisms and small arthropods, such as mites and collembolans, are of much greater importance in soils.

Trophic complexity

One particularly surprising result of analysis of arthropod community struc- ture in trees is that the division of arthropod species between feeding guilds appears to be remarkably similar in temperate and tropical samples (Stork, 1987). Furthermore, the division of species is remarkably constant for some guilds across trees in the same set of samples (Moran and Southwood, 1982; Stork, 1987). This result would suggest that, even though the proportions of different taxonomic groups in particular ecosystems vary in different parts of the globe, the trophic structure of such ecosystems remains relatively con- stant, at least at the species level. Such global patterns are being examined to assess this further, as confirmation of such patterns will add confidence to extrapolations of global species numbers estimates from food-web ratios. It will also provide a yardstick for investigation of guild structure in managed systems. It may prove that departures from the pattern are associated with instability and major pest outbreaks (perhaps in relation to insecticide application) through elimination of key components of the ecosystem.

All animal species are dependent either directly or indirectly on the resource base provided by plant material. This resource shows marked biochemical heterogeneity when living, a heterogeneity that declines as it dies and is decomposed. This heterogeneity is reflected positively in the diversity of the defoliator, bark feeding and decomposer guilds in South-East Asian rain forest. The relationship of herbivore diversity to floristic diversity is highly complex and not always in close parallel (Holloway, 1989), but an understanding of the extent of specialism and of coevolutionary pattern through comparisons of the systematics of plant groups and their herbivore predators (e.g. Mitter and Farrell, 1991) is essential for refining a major extrapolative step in Erwin's (1982) global species estimate. It will also enable predictions of reduction of overall diversity of organisms in floristically simplified managed systems to be made from a reliable information base.

Invertebrates as environmental indicators

We have so far presented a sketch of salient features of the natural profile of biodiversity. With current concern that our activities are rapidly changing the structure and functioning of global ecosystems, it is necessary to assess and monitor natural or human-induced changes in that profile.

The causes of perturbation are often clear: measurement of biological variables gives a direct indication of the *effects* of perturbation. Such variables must show a prompt and accurate response to perturbation and must reflect some aspect of the functioning of the ecosystem. They must be readily and economically assessed if a regular programme of monitoring is desired. The organisms used should be universal in distribution yet show individual specificity to temporal or spatial pattern in the environment. These specific tolerances should be measurable to ensure the monitoring system includes an element of predictiveness. Will a biological control organism establish in a particular climate belt? Will a pest be likely to establish in a given area if there is a given shift in climatic parameters?

Invertebrates have many qualities that render them suitable as indicator groups: generality of distribution, trophic versatility, specialism within that generality and versatility, rapid response to perturbation, taxonomic tractability, statistically significant abundance, and ease of sampling (Holloway, 1983; Hammond, 1990). A selection of indicator groups are needed for a full assessment of biodiversity and our impact on it. Proven, standardized sampling techniques are already available (Hammond, 1990) as are proven, standardized statistical methods, at least for marine sampling (e.g. Food and

Agriculture Organization *et al.*, 1984). For the monitoring of pollution and disturbance in marine environments such assessments, through the use of nematodes, are already in operation (Lambshead *et al.*, 1983).

A number of examples of how invertebrates respond to major perturbations have been given above. Here we give further examples from light-trap data for night-flying moths, comparing facets of biodiversity in natural forest and managed systems. Moths in their larval stage are mainly foliage-feeding, but other guilds are present: detritivores of both plant and animal material; fruit, flower and seed predators; stem and timber borers; lichen and algal browsers; insectivores. They fulfil all the criteria listed above for a good indicator group (Holloway, 1983, 1984*b*, 1985).

Moth diversity in the natural forests of the Indo-Australian tropics has a humped distribution with respect to altitude (Holloway, 1987, 1991; Holloway *et al.*, 1990). Figures 5.8 *a* and *b* illustrate such profiles of natural diversity from Sulawesi and Seram, respectively. The open circles in each represent samples from cultivated areas. In Sulawesi the cultivation samples are from field-crops or plantations. Moth diversity is effectively halved, probably more so as most samples include stray or relictual forest components.

This loss of diversity is differential among moth families, as can be seen from the proportions in forest versus agricultural samples from the same altitude (Fig. 5.8*c*).

Families with a predominance of arboreal-feeding larvae, such as Geometridae, Notodontidae and Lymantriidae, decline or disappear, leaving the field to groups such as the trifine subfamilies of Noctuidae and the Macroglossinae of the Sphingidae, where the larvae feed on herbaceous plants. These groups include most of the biogeographically widespread pest species mentioned earlier. In cluster analyses of the Sulawesi data (trifine subfamilies of Noctuidae; Arctiidae; ennomine Geometridae), this was also apparent (Holloway *et al.*, 1990). For the trifine noctuids the samples from agricultural habitats formed a strong cluster, distinguished from lowland and montane forest clusters. This effect is much weaker for the Arctiidae, with fewer open habitat specialists, and virtually absent for the Ennominae where arboreal feeding is prevalent.

On the Moluccan island of Seram, it was possible to compare a natural forest transect with both modern agriculture and a long-standing shifting cultivation system (Holloway, 1991). The natural profile with altitude is a typical humped curve (Fig. 5.8*b*). The modern agricultural systems (at Wahai) were at low altitude and exhibited a drop in diversity and changes in family proportions much as did those in Sulawesi, though the samples were small owing to a dry season more acute than usual.

The shifting cultivation area was at middle altitudes in a forest-enclosed valley enclave in the middle of the island (Manuselah). The villages are surrounded by a mosaic of habitats representing a succession from freshly

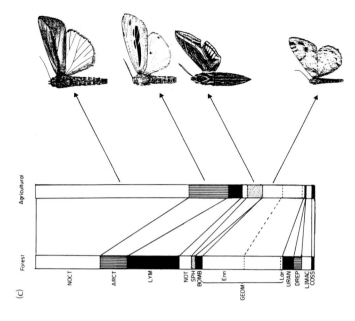

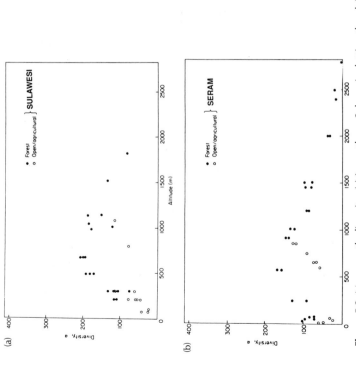

Figure 5.8. Macromoth diversity. (a) In northern Sulawesi, plotting the alpha-statistic against altitude; ● = samples from natural forest; O = samples from agricultural systems (after Holloway, 1987). (b) In Seram, Moluccas, axes as in (a); 'o' = at low altitudes indicates two samples from coconut plantations (lowest diversity) and two from riverine localities with some human disturbance, those at middle altitudes are from a shifting cultivation cycle: the three lowest values are from recently cleared garden areas, the next highest is from a recently overgrown area and the two upper ones are from gardens abandoned for several decades (from Holloway 1991). (c) Change in the proportions of different moth families that accompanies the loss of diversity on conversion of forest to agriculture in (a); the samples are from forest and open fields in close proximity at about 200 m in the Dumoga Plain; proportions of Noctuidae, Arctiidae and macroglossine Sphingidae increase relatively, while other families such as Geometridae and Lymantriidae show a sharp decline; family abbreviations as in Fig. 5.2 (moth portraits by B. Hargreaves).

cleared and burnt areas, through cultivated gardens, to stages of forest regeneration with giant bamboo giving way to more diverse secondary forest. Moth diversity is low in freshly cleared and cultivated areas (the three points at lowest altitude) but rises as forest regenerates virtually to the level of that in undisturbed forest (the higher altitude sites where farmland has been abandoned for several decades). The moths characteristic of this succession are not open habitat specialists, the field-crop pests, but are early gap phase successional species of the natural forest cycle. Family proportions resemble those of the natural forest at all stages of the cycle.

This suggests that a sustainable cycle of shifting cultivation, or perhaps some system of forest farming or agroforestry such as that described for Sumatra by Michon *et al.* (1986), may be the mode of human exploitation of tropical ecosystems most compatible with preservation of high biological diversity. The closest approach to agroforestry sampled in Sulawesi, the coconut/clove system, supported higher moth diversity and a more 'forest-like' profile of family proportions than did open systems or pure coconut plantation (Holloway, 1987).

Much more sampling of a wider range of indicator groups in managed systems is needed to enable us to gauge the capacity of each to conserve biodiversity. Groups containing a high proportion of specialist herbivores, such as insects, will provide a much more sensitive assessment of effects than will less specialist groups such as vertebrates that respond to the structural dimensions of the ecosystem rather than to floristic heterogeneity.

Invertebrates are well-known as indicators of past climate change (e.g. Coope, 1975) and of antiquity of forests (Terrell-Nield, 1990; Harding and Rose, 1987). We suggest that they may also be used to monitor the effects of future climate change. For example, the open habitat specialist community of moths on Norfolk Island has demonstrated this by its response to two abnormally dry years in a period of regular sampling of over 10 years (Holloway, 1977, 1983). Plots of the logarithm of species abundance against species rank order (Fig. 5.9) for the 24 most abundant species show a more even distribution of abundance in a year where rainfall was significantly below average for each month than in the 2 years of normal rainfall before and after. This effect was differential among the species, with the two major pasture-feeding armyworms changing places in the rankings: the savanna habitat *Mythimna* is favoured in the dry year, and the humid tropical *Spodoptera* in the two wetter ones. Study of the proportions of these species over a long sequence of years has shown that this response is repetitive (Holloway, 1983).

Data such as these may help guide us in assessing the potential effects of global warming.

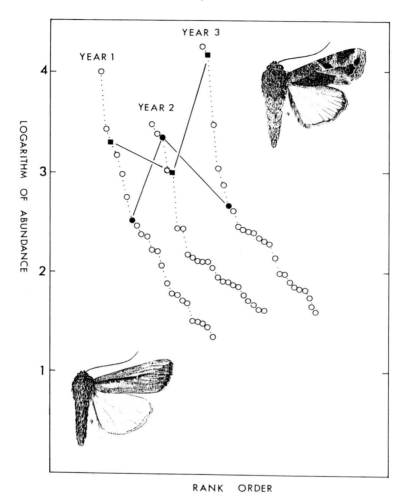

Figure 5.9. Species/abundance curves for the most abundant 23 macromoth species of the Norfolk Island open habitat moth community for 2 years of normal rainfall on either side of a year where the rainfall for each month was considerably less than the average. Abundance is more evenly distributed among the species in the dry year, and there were rank order changes among the species. This is illustrated by two pasture-feeding armyworms: *Spodoptera mauritia* (top right, solid squares), a humid tropical species, predominates in the wetter years, and *Mythimna loreyimima* (bottom left, solid circles), a savanna species, predominates in the dry year (after Holloway, 1977; moth portraits by G. duHeaume).

Conclusions

We have surveyed what is known of the parameters that define 'biodiversity', highlighting ways of how invertebrates can be used to understand it, both within the structure of single ecosystems and with respect to large-scale and global patterns. Such understanding is vital as it is evident that our activities can cause the diminution of all dimensions of biodiversity. Only through a full understanding of the natural determinants of biodiversity can we hope to minimize its shrinkage. We have shown that invertebrates, and in particular insects, are a significant component of faunal diversity that cannot be ignored in the assessment of biodiversity – it is not enough just to examine more obvious groups such as plants, birds and mammals. Having reviewed the problem we make a number of recommendations for future action. We see the priorities for maintaining biodiversity through sustainable agriculture, in its widest sense, as the following.

1. We need to clarify all dimensions of biodiversity to gain a clearer picture of its natural profile.
2. We need to assess the changes that occur when we modify or attempt to manage this profile.
3. We recommend the adoption of a standard set of sampling methods or a 'sampling package' and methods of analysis so that results can be compared in different parts of the world.
4. We need to use tried and tested indicator groups on a 'horses for courses' basis. Groups such as moths, ground beetles, ichneumonids, mites, ants and termites may be suitable but further careful consideration of other groups is required (see Hammond, 1990).

These tasks can only be achieved with the full involvement of a critical mass of specialists in systematics supported by broad-based reference collections of international significance.

Finally, too often the loss (and rarely gain) of biodiversity is omitted from the economic equations produced by agronomists when weighing up the pros and cons of various agricultural schemes. Too often valuation of biodiversity – or loss of it – is carried out retrospectively. Auditing of biodiversity will be difficult and will vary locally and regionally in relation to other economic and social criteria. It is time that the economist, conservationist, farmer, forester and biologist sat down together.

References

Ackery, P.R. and Vane-Wright, R.I. (1984) *Milkweed Butterflies, their Cladistics and Biology.* British Museum (Natural History), London.

Adis, J., Lubin, Y.D. and Montgomery, G.G. (1984) Arthropods from the canopy of inundated and terra firme forests near Manaus, Brazil, with critical considerations of the Pyrethrum-fogging technique. *Studies on Neotropical Fauna and the Environment* 19, 223–236.

Adis, J. and Schubart, H.O.R. (1985) Ecological research on arthropods in central Amazonian forest ecosystems with recommendations for study procedures. In: Cooley, J.H. and Golley, F.B. (eds), *Trends in Ecological Research for the 1980's. Nato Conference Series, Series 1: Ecology.* Plenum Press, London, New York, pp. 111–144.

Barnes, R.D. (1989) Diversity of organisms: How much do we know? *American Zoologist* 29, 1075–1084.

Cock, M.J.W., Godfray, H.C.J. and Holloway, J.D. [eds] (1987) *Slug and Nettle Caterpillars. The Biology, Taxonomy and Control of the Limacodidae of Economic Importance on Palms in South-east Asia.* CAB International, Wallingford.

Collins, N.M. and Morris, M.G. (1985) *Threatened Swallowtail Butterflies of the World, the IUCN Red Data Book.* International Union for the Conservation of Nature, Gland and Cambridge.

Coope, G.R. (1975) Climatic fluctuations in northwest Europe since the Last Interglacial, indicated by fossil assemblages of Coleoptera. In: Wright, A.E. and Moseley, F. (eds), Ice Ages: Ancient and Modern. *Geological Journal, Special Issue* 6 153–168.

Croizat, L. (1964) *Space, Time, Form: The Biological Synthesis.* L. Croizat, Caracas.

Di Castri, F. and Younès, T. [eds] (1990) Ecosystem function of biological diversity. *Biology International, Special issue* 22, 1–20.

Diamond, J.M. (1985) How many unknown species are yet to be discovered? *Nature,* 315, 538–539.

Diamond, J.M. and May, R.M. (1981) Island biogeography and the design of natural reserves. In: May, R.M. (ed.), *Theoretical Ecology, Principles and Applications.* 2nd ed. Blackwell Scientific Publications, London, pp. 228–252.

Erwin, T.L. (1982) Tropical forests: their richness in Coleoptera and other arthropod species. *Coleopterist's Bulletin* 36, 74–75.

Erwin, T.L. (1983) Beetles and other insects of tropical forest canopies at Manaus, Brazil, sampled by insecticidal fogging. In: Sutton, S.L., Whitmore, T.C. and Chadwick, A.C. (eds), *Tropical Rain Forest Ecology and Management.* Blackwell Scientific Publications, Oxford, pp. 59–75.

Food and Agriculture Organization, International Organization for Conservation, and United Nations Environment Programme (1989) *Lecture Notes prepared for the Training Workshop on the Statistical Treatment and Interpretation of Marine Community Data.* FAO, Athens.

Gauld, I.D. (1986) Latitudinal gradients in ichneumonid species richness in Australia. *Ecological Entomology* 11, 155–161.

Gilbert, F.S. (1980) The equilibrium theory of island biogeography: fact or fiction. *Journal of Biogeography* 7, 209–235.

Grassle, J.F. (1989) Species diversity in deep-sea communities. *TREE* 4, 12–15.

Hammond, P.M. (1990) Insect abundance and diversity in the Dumoga-Bone National Park, North Sulawesi, with special reference to the beetle fauna of lowland rainforest in the Toraut region. In: Knight, W.J. and Holloway, J.D. (eds), *Insects and the Rain Forests of South East Asia (Wallacea)*. The Royal Entomological Society of London, London, pp. 197–254.

Harding, P.T. and Rose, F. (1986) *Pasture-woodlands in lowland Britain. A Review of their Importance for Wildlife Conservation*. Institute of Terrestrial Ecology, Huntington.

Hodkinson, I.D. and Casson, D. (1991) A lesser predilection for bugs. Hemiptera (insecta) diversity in tropical rain forests. *Biological Journal of the Linnean Society* 43, 101–109.

Holloway, J.D. (1969) A numerical investigation of the biogeography of the butterfly fauna of India, and its relation to continental drift. *Biological Journal of the Linnean Society* 1, 373–385.

Holloway, J.D. (1973) The taxonomy of four groups of butterflies (Lepidoptera: Rhopalocera) in relation to general butterfly distribution in the Indo-Australian area. *Transactions of the Royal Entomological Society of London* 125, 125–176.

Holloway, J.D. (1974) The biogeography of Indian butterflies. In: Mani, M.S. (ed.), *Ecology and Biogeography in India*. W. Junk, The Hague, pp. 473–499.

Holloway, J.D. (1977) *The Lepidoptera of Norfolk Island, their Biogeography and Ecology*. [Series Entomologica No. 13.] W. Junk, The Hague.

Holloway, J.D. (1979) *A Survey of the Lepidoptera. Biogeography and Ecology of New Caledonia*. [Series Entomologica No. 15.] W. Junk, The Hague.

Holloway, J.D. (1983) Insect surveys – an approach to environmental monitoring. *Atti XII Congresso Nazionale Italiano di Entomologia, Roma 1980*, 239–261.

Holloway, J.D. (1984a) The IUCN Invertebrate Red Data Book: a challenge for biogeography. *Journal of Biogeography* 11, 175–180.

Holloway, J.D. (1984b) The larger moths of Gunung Mulu National Park: a preliminary assessment of their distribution, ecology and potential as environmental indicators. *Sarawak Museum Journal* 30 (*Special Issue* 2), 149–190.

Holloway, J.D. (1985) Moths as indicator organisms for categorising rain forest and monitoring changes and regeneration processes. In: Chadwick, A.C. and Sutton, S.L. (eds), *Tropical Rain-Forest*. Leeds Philosophical and Literary Society, Leeds, pp. 235–242.

Holloway, J.D. (1986) Lepidoptera fauna of high mountains in the Indo-Australian tropics. In: Vuilleumier, F. and Monasterio, M. (eds), *High Altitude Tropical Biogeography*. Oxford University Press, New York, pp. 533–556.

Holloway, J.D. (1987) Macrolepidoptera diversity in the Indo-Australian tropics: geographic, biotopic and taxonomic variations. *Biological Journal of the Linnean Society* 30, 325–341.

Holloway, J.D. (1989) Moths. In: Lieth, H. and Werger, M.J.A. (eds), *Tropical Rain Forest Ecosystems, Biogeographical and Ecological Studies*. [Ecosystems of the World No. 14B.] Elsevier, Amsterdam, pp. 437–453.

Holloway, J.D. (1991) Aspects of the biogeography and ecology of the Seram moth fauna. In: Edwards, I. and Proctor, J. (eds), *The Natural History of Seram*, in press.

Holloway, J.D. and Jardine, N. (1968) Two approaches to zoogeography: a study based on the distributions of butterflies, birds and bats in the Indo-Australian area. *Proceedings of the Linnean Society of London*, 179, 153–188.

Holloway, J.D., Robinson, G.S. and Tuck, K.R. (1990) Zonation in the Lepidoptera of northern Sulawesi. In: Knight, W.J. and Holloway, J.D. (eds), *Insects and the*

Rain Forests of South East Asia (Wallacea). The Royal Entomological Society of London, London, pp. 153–166.

International Union for Conservation of Nature and Natural Resources (1987) *Centres of Plant Diversity. A Guide and Strategy for their Conservation.* International Union for Conservation of Nature and Natural Resources, London.

International Union for Conservation of Nature and Natural Resources, United Nations Environment Programme and World Resources Institute (1990) *The Biodiversity Conservation Strategy Programme, Mobilizing Worldwide Action to Sustain the Living Resources of our Planet.*

Lambshead, P.J.D., Platt, H.M. and Shaw, K.M. (1983) The detection of differences among assemblages of marine benthic species based on an assessment of dominance and diversity. *Journal of Natural History* 17, 859–874.

MacArthur, R.H. and Wilson, E.O. (1967) *The Theory of Island Biogeography.* Princeton University Press, Princeton.

May, R.M. (1981) Patterns in multi-species communities. In: May, R.M. (ed.), *Theoretical Ecology, Principles and Applications.* 2nd ed. Blackwell Scientific Publications, London, pp. 197–227.

May, R.M. (1986) How many species are there? *Nature*, 324, 514–515.

May, R.M. (1988) How many species are there on earth? *Science*, 241, 1441–1449.

Michon, G., May, F. and Bompard, J. (1986) Multistoried agroforestry garden system in West Sumatra, Indonesia. *Agroforestry Systems* 4, 315–338.

Mitter, C. and Farrell, B. (1991) Macroevolutionary aspects of insect/plant relationships. In: Bernays, E.A. (ed.), *Insect/Plant Interactions.* Vol. 3. CRC Press, Boca Raton, in press.

Moran, V.C. and Southwood, T.R.E. (1982) The guild composition of arthropod communities in trees. *Journal of Animal Ecology* 51, 289–306.

Morse, D.R., Lawton, J.H., Dobson, M.M. and Williamson, M.H. (1985) Fractal dimension of vegetation and the distribution of arthropod body lengths. *Nature* 314, 731–733.

Morse, D.R., Stork, N.E. and Lawton, J.H. (1988) Species number, species abundance and body–length relationships of arboreal beetles in Bornean lowland rain forest trees. *Ecological Entomology* 13, 25–37.

Myers, N. (1988) Threatened biotas: 'Hot spots' in tropical forests. *The Environmentalist* 8, 187–208.

Myers, N. (1991) The biodiversity challenge: expanded hot-spot analysis. *The Environmentalist* 10, in press.

Nelson, G. (1989) Species and taxa: systematics and evolution. In: Otte, D. and Endler, J.A. (eds), *Speciation and Its Consequences.* Sinauer, Sunderland, Massachusetts, pp. 60–81.

Pearson, D.L. and Ghorpade, K. (1989) Geographical distribution and ecological history of tiger beetles (Coleoptera: Cicindelinae) of the Indian subcontinent. *Journal of Biogeography* 16, 333–344.

Preston, F.W. (1962) The canonical distribution of commonness and rarity. *Ecology 39, 185–215, 410–432.*

Robinson, G.S. (1975) Macrolepidoptera of Fiji and Rotuma, a Taxonomic and Geographic Study. E.W. Classey, Faringdon.

Stork, N.E. (1987) Guild structure of arthropods from Bornean rain forest trees. *Ecological Entomology* 12, 69–80.

Stork, N.E. (1988) Insect diversity: facts, fiction and speculation. *Biological Journal of the Linnean Society* 35, 321–337.

Stork, N.E. (1991) The composition of the arthropod fauna of Bornean lowland rain forest trees. *Journal of Tropical Ecology* 7, 161–180.

Stork, N.E. and Brendell, M.J.D. (1990) Variation in the insect fauna of Sulawesi trees with season, altitude and forest type. In: Knight, W.J. and Holloway, J.D. (eds), *Insects and the Rain Forests of South East Asia (Wallacea)*. The Royal Entomological Society of London, London, pp. 173–190.

Stork, N.E. and Brendell, M.J.D. (1991) Arthropod diversity studies in lowland rain forest of Seram: Indonesia. In: Edwards, I. and Proctor, J. (eds), *The Natural History of Seram*, in press.

Stork, N.E. and Gaston, K.J. (1990) Counting species one by one. *New Scientist* 127 (1729), 43–47.

Strong, D.R., Lawton, J.H. and Southwood, T.R.E. (1984) *Insects on Plants: Community Patterns and Mechanisms*. Blackwell Scientific Publications, Oxford.

Sugihara, G. (1980) Minimal community structure: an explanation of species abundance patterns. *American Naturalist* 116, 770–787.

Swift, M.J. and Anderson, J.M. (1989) Decomposition. In: Lieth, H. and Werger, M.J.A. (eds), *Tropical Rain Forest Ecosystems, Biogeographical and Ecological Studies*. [Ecosystems of the World No. 14B.] Elsevier, Amsterdam, pp. 547–569.

Terrell-Nield, C. (1990) Is it possible to age woodlands on the basis of their carabid beetle diversity? *The Entomologist* 109, 136–145.

Thomas, C. (1990) Fewer species. *Nature* 347, 237.

Vane-Wright, R.I. (1991) Transcending the Wallace line: do the western edges of the Australian Region and the Australian Plate coincide? *Australian Systematic Botany*, in press.

Vane-Wright, R.I., Humphries, C.J. and Williams, P.H. (1991) What to protect? Systematics and the agony of choice. *Biological Conservation* 55, 235–254.

Wilson, E.O. (1988) The current state of biological diversity. In: Wilson, E.O. (ed.), *Biodiversity*. National Academy Press, Washington, DC, pp. 3–18.

Wolf, E.C. (1987) *On the Brink of Extinction: Conserving the Diversity of Life.* [Worldwatch Paper No. 78.] Worldwatch Institute, Washington DC.

Discussion

Punithalingam: Has the selected collecting of Lepidoptera for commercial purposes in South-East Asia had any influence on biodiversity, and will this practice deplete populations of rare species?

Holloway: On the whole, I think not. Indeed I can cite an instance where a conservationist in a natural forest in the Philippines was offered some bufferfly specimens. He declined, but if he had taken the opportunity to purchase some and point out to the collector how these and so his livelihood could be endangered if the natural forest was reduced, that would have been a more positive step for conservation.

Greenland: You drew attention to differences in biodiversity with altitude and contrasted these with stages of shifting cultivation from cultivated to redeveloped forest. Could you give more information about biodiversity relating to the different steps in the shifting cultivation cycle?

Holloway: This was a pilot study. The villagers clear among trees which they use and plant bananas, cassava, etc., between them. After a few seasons they move and the forest returns. At first the moth diversity declines, but then it increases, fern-feeders come in, and gradually others, and the level of diversity starts to build up again. There is a need for more monitoring experiments of this kind in areas where the insect fauna is well-known.

Six

General Discussion: Session I

Session Chairman: M.H. Arnold, *Hamlec, 4 Shelford Road, Whittlesford, Cambridgeshire CB2 4PG, UK*

Session Rapporteur: P.R. Scott, *CAB International, Wallingford, Oxfordshire OX10 8DE, UK*

Introduction

The theme of this session is the importance of microorganisms and invertebrates as components of biodiversity. In our discussion we have to think about priorities, what we mean by biodiversity, and address the question as to the value of biodiversity in relation to sustainability and food production. The importance of conservation in collections as opposed to *in situ* conservation or 'setaside' has also to be considered.

Discussion

Rothschild: How can the issue of setting priorities for work in biodiversity be resolved? There is a need to be selective in identifying the key targets for work, but very often the key areas can only be recognized when the research has already been undertaken. Such research is long term and time is not on our side; unique habitats are being destroyed at an alarming rate. One is faced with a dilemma if resources are limited. On the one hand the systems approach requires the identification of key elements of the fauna and flora that play a role in different ecosystems. On the other, there is the need to conserve pristine areas with inherently greater biodiversity even though the importance of the different elements of that biodiversity is unknown. Perhaps both areas need to be equally funded, or could a formula for suggesting how each should be supported be found?

Kapoor: As resources are limited, the pressing requirement of developing countries is to identify the key species that are going to be of importance for economic development. Ethnobiological data are especially important in recognizing species that are socially useful.

The Biodiversity of Microorganisms and Invertebrates: Its Role in Sustainable Agriculture.
Edited by D.L. Hawksworth. © CAB International 1991.

Hulse: One issue touched on has been the shortage of taxonomists; they have been referred to as a dying breed. We need to consider what is the position with regard to the people who are to carry out the work required. Training courses such as those organized by CAB International help but many more resources are surely required. Is the first crisis a people crisis?

Gamez: Costa Rica's National Biodiversity Inventory has had to face just this problem. We may have 0.5 million species and very few taxonomists. This is a broad participatory project being conducted in large measure by lay workers trained for the task, 'parataxonomists', working in close collaboration with both national and international curators and taxonomists. Specimens and other information from the parataxonomists flow into the National Biodiversity Institute (INBio) where they are processed into the National Biodiversity Collections and National Biodiversity Database. This information then moves into a National Dissemination and Extension Service which provides information to users in appropriate formats.

Perfect: A major concern is the consideration of functionality. If we use a single group of organisms as indicators, can we be sure all functions are represented in it? The particular phenotypes that carry the molecular basis for functionality may not be the most useful way of describing biodiversity. We must look beyond traditional techniques.

Tinker: The use of less-trained staff in survey work is an interesting approach, and has also been employed in the Rothamsted Insect Survey. Concern is being expressed about the shortage of taxonomists in relation to the needs of biodiversity. Indeed, I receive continual suggestions that something should be done to support taxonomy. On probing, these turn out to mean at least three different things: (i) the lack of support for taxonomic research; (ii) the lack of short-term vocational training for taxonomists; and (iii) the poor level of whole-organism training in undergraduate degree courses. We need to identify precisely where the problems lie; my view is that the third point may be the most serious.

Dhiman: I find there is a great diversity in the conceptual meanings of biodiversity. Most of the discussion has involved genetic and species diversity in natural ecosystems. This is a long-term context. I consider that the strategic deployment of crop varieties in relation to time and space to protect crops such as cereals from diseases such as rusts and smuts, of biocontrol agents, and of plant growth promoting organisms in crop soils, should be viewed as short-term biodiversity tactics.

Burns: I suggest that in some areas of the microbial world diversity is increasing rather than decreasing. By diversity I mean biochemical or metabolic diversity as it is what microorganisms do, or can be persuaded to do, that is important and not what they are. Microbes have been exposed to a whole range of 'exotic' agrochemicals and other industrial compounds. The microbial world has responded rapidly and effectively to this challenge by developing novel metabolic pathways to either detoxify the chemicals or, most commonly, to utilize the compounds as carbon, nitrogen, or energy sources. It would be of interest to discover the comparative rates of diversity losses and gains in microbial communities.

Morris: Priorities are set by the current generation. Sustainability means keeping options open for future generations. Future needs are likely to be unpredictable, and biodiversity needs to be conserved because of this unpredictability. Preservation only on the basis of current utility is risky. The need to sustain biodiversity for future generations is particularly important to stress to policy-makers who are not accustomed to thinking of long-term benefits.

Hadley: Biological diversity is a notion that has entered the arena of public and political concern. Indications of this are references in the reports of the Bruntland Commission, the Commonwealth Heads of Government Meeting in Kuala Lumpur, and the Summit of the Arch Meeting in Paris. The entry of biological diversity into the public domain represents an opportunity but also a responsibility for biologists, encapsulated in being able to answer the question 'So what?' Being able to provide insights into the functional significance of biodiversity calls for contributions from a wide range of disciplines, both within and outside of biology. It also calls for the adoption of some agreed research designs and protocols, whereby a group of scientists agree to address one or more research questions using a shared or compatible methodology over a predetermined period. In working towards collaborative studies which 'compare the comparable' in ecology in general, and biological diversity in particular, I was interested in Dr Holloway and Dr Stork's (Chapter 5) proposals to develop and adopt 'standard sampling packages' and 'horses for courses' indicators and indices. Further progress on these lines would be of considerable interest.

Platt: I have been involved in UNESCO-funded GEEP (Group of Experts on Environmental Pollution) workshops in Oslo, Bermuda, and Bremerhaven aimed at assessing diversity in the marine environment. The aim is to produce a manual of 'approved' workable and cost-effective techniques based on a series of levels: molecular/physiological, species level, and community level. It might be timely to adopt a similar approach with regard to the biodiversity and environmental quality of agricultural soils.

Samways: There are two fundamentally different philosophical foundations to biodiversity: pragmatic/economic, and deep ecological/bioempathetic. Both are recognized and significant, especially for future generations of man, *and* other biota. Both approaches lead down the same road, to habitat preservation and conservation, where organisms exist in their own right and also remain as a pool for utilization.

Altieri: This meeting also needs to consider what sustainable agriculture is. The concept of sustainability has to be broad and not only environmentally sound, but socially just, environmentally viable, culturally acceptable, and politically appropriate. Much of the biodiversity is in the Third World and countries there need to be compensated for uses made of their biodiversity in order to promote their growth, for example as biocontrol agents. The conservation of biodiversity has to be seen in political and economic terms.

Khalil: This session has concentrated on insects and microorganisms, but I would like to draw attention to nematodes, which are among the most abundant animals on earth. Although more species of insects have been described, many have been found to harbour at least one or two species of parasitic nematodes. Most vertebrates and some other invertebrate groups are parasitized by several species of nematodes, and there are also species confined to particular plants. Further, the species of free-living marine, freshwater, and soil nematodes probably far outnumber those that are parasitic. Where nematodes occur they are often in very large numbers, for example more than 90 000 were found in a single rotting apple, and 3–9 billion per acre may be found in good farmland in the USA. Nematodes have a great diversity in morphology, life-cycles and behaviour, some causing diseases of extreme importance to humans as well as to domesticated and wild animals and to crops. They also are good bioindicators of environmental pollution, and have been used extensively in cytological and ageing studies.

Heydon: Financial resources are limited, and with regard to 'keeping options open' it is necessary to determine economic priorities as between devoting these to either (i) maintaining a healthy taxonomic research capability which can be used to respond to pressing topical concerns in the long term, or (ii) undertaking research perceived to be economically viable and to have short-term benefits based on the exploitation of beneficial organisms. Given the current importance of biodiversity for sustainable agriculture, it is vital that a healthy research capability be preserved for the future, and this in turn will require appropriate political recognition and support.

Gamez: We must not forget that people's needs and aspirations must be central concerns in our plans and programmes to preserve and use sustainable biodiversity. There is a need to develop local capacities and strong local organizations in tropical countries, where most of the biodiversity is concentrated. In Costa Rica the conservation programme is based on the premise that only that biodiversity that is intellectually and economically embedded in the society that contains it will survive. We believe we have 10 years to accomplish that goal, and so have adopted the following strategy: (i) save it (in national parks); (ii) know it; and (iii) put it to use (for society, economically, or intellectually). Saving tropical diversity must move from being a topic for extratropical discussion to one of daily concern in local tropical societies. This requires economic and intellectual inputs from developed countries, and a change of attitudes and priorities in the way the tropics are perceived, from a resource to be used to being a partner in the protection and sustainable use of tropical biodiversity.

Yusof Hashim: The preservation or conservation of biodiversity has to be seen in a proper perspective and balance. In countries where food is plentiful and modern agriculture well-established such a concern could be addressed with appropriate resources, both expertise and funds. In contrast in less-developed countries, where food is insufficient to feed ever increasing populations, food production at whatever cost is of the utmost concern. Should not there be a balanced approach or rationale between biodiversity conservation and food/crop production so that the less-developed countries are not disadvantaged? Some kind of model system needs to be developed and studied.

Mohamed Senawi: There is a need for a correct perspective between biodiversity and a balanced ecosystem, particularly in the context of agricultural development. In a major commodity-producing country such as Malaysia, intensive plantation cropping brings higher productivity per unit area, and concurrently a higher amount of agricultural waste per unit area. Natural biodegradation processes are unable to recycle these biomasses and environmental hazards result. Biotechnology provides possibilities for the recycling of agricultural wastes into useful products such as food, feeds, and biofertilizers through solid fermentation. Unfortunately, progress in this field is hampered by inadequate information on microorganisms able to break down lignocelluloses, and organizations such as CAB International are looked to for advice.

Szabolcs: Biodiversity and biota change not only in space but in time. The time changes are accelerated by anthropogenic effects, and often are negative, not only in tropical countries but also in arid regions. For example, the Aral Sea and its surroundings have salinized during the last 20 years and the biota has changed radically. I also wish to stress that there are two different approaches to biodiversity studies: (i) environment orientated; and (ii) production orientated. The gap between these needs to be bridged.

Robinson: The need for taxonomic expertise in tackling even the simplest questions of biodiversity has been alluded to. The basis for any informed statements on biodiversity must be a broad, consistently competent level of expertise based on well-maintained reference collections. Such a facility requires continuous and adequate long-term core funding. Centres of excellence are not maintained by short-term contract money, and governments of developed countries should be encouraged to continue and improve core funding for taxonomy. Given the continuance of the primary international biosystematics centres, as pointed out by Dr Gamez there is a need to transfer taxonomic expertise and information by training, scientific exchange, and the setting up of regional reference collections and libraries. This comes within the purview of international and regional funding agencies, to whom a plea for substantial financial backing to provide for the transfer of 'taxonomic technology' to the developing world should be made.

Session Chairman's conclusions

Several speakers alluded in one way or another to the robustness or resilience of biological systems, especially those based on microorganisms, and to the fact that organisms respond to changes in the environment, so there is perhaps no need to attempt to preserve existing biodiversity exactly as it is.

Considerations such as these led to two recurring questions that arose both implicitly and explicitly during the discussions: (i) Why we should be concerned about the preservation of biodiversity in microorganisms and invertebrates? (ii) Assuming that we are concerned, what should we do about it?

Why should we be concerned?

The point was made that the world community has been concerned for some time about the preservation of diversity in higher plants and vertebrates, and that there seem to be no logical reasons for *not* giving equal attention to microorganisms and invertebrates, particularly taking into account their importance in such processes as biological control and nutrient recycling. There is also the need to maintain the capacity of production systems to change in the face of changing pressures, such as population pressures or climatic change, and the associated need to preserve the role of microorganisms and invertebrates in facilitating such changes.

The argument was also made, one it is very difficult to counter, that we do not know what will be required in the future and therefore we have a responsibility to future generations to preserve as much as we reasonably can.

In other words, there was a clear consensus that we should persuade policy-makers to be more concerned than they currently are about the conservation of diversity in microorganisms and invertebrates.

Given the need, what should we do about it?

There seemed to be general support for the notion of what Professor Stewart (Chapter 1) called a 'global setaside scheme'. That is, a globally co-ordinated scheme to conserve representative samples of different ecosystems. This immediately gives rise to a whole list of questions, the answers to which can be given only through further research directed towards the *analysis* of biodiversity. Here, we encountered some dichotomy in the views expressed, whether biodiversity is best analysed in terms of taxonomy (i.e. the recognition of species), or in terms of variations in gene functions within populations of organisms.

The suggestion was made that the tools of molecular biology might be used to make systematic comparisons of populations of organisms in different ecosystems and to reduce the need for traditional taxonomic work. Others saw traditional taxonomy as *crucial*, being the only satisfactory framework on which information systems could be built. In this view, the key to information exchange is rapid and accurate identification. It requires reasonably comprehensive collections and accumulation of biological data associated with them. In so far as a consensus emerged, it was that these two approaches, the molecular and the taxonomic, are by no means mutually exclusive but are, in reality, highly complementary. Consequently, both approaches should continue to be supported.

Other points brought out by various speakers related to the need for standard methodologies and agreed protocols in surveying biodiversity to give greater comparability of work in different regions and on different ecosystems. Also, the need for further research to test the validity of using indicator groups to monitor *changes* in biodiversity was raised.

There seemed to be general agreement that the primary needs are:

1. To find out more about the distribution of natural diversity in micro-organisms and invertebrates.
2. To find out more about how the diversity is affected under a range of managed production systems.
3. To use the results of such research to give greater inputs into the requirements for sustainable production systems and to provide a more quantitative basis for making decisions on what to conserve, where to conserve it and how to conserve it, including the more precise definition of the respective rules of collections and *in situ* conservation.

In other words, we cannot expect to give definitive answers to many of the questions until we know more about the nature and distribution of diversity in these organisms.

Pervading all the presentations in Session I, as well as the Discussions from the floor, was an emphasis on the need:

1. To incorporate an awareness of biodiversity into all levels of education from primary school to university.
2. The need for university training in *biosystematics*, using that term to embody both techniques in molecular biology and the principles of taxonomy.

3. To involve farmers and extension workers in the whole process of increasing awareness of biodiversity and its implications.

4. The need to strengthen taxonomic and biosystematic services at the national level, especially in developing countries.

II
Importance of Biodiversity in Sustaining Soil Productivity

Seven

The Diversity of Soil Organisms

K.E. Lee, *CSIRO Division of Soils, Private Mail Bag No. 2, Glen Osmond, South Australia 5064, Australia*

ABSTRACT: The biodiversity of soils is considered, particularly in relation to the larger invertebrates able to reshape soil to suit their requirements. Following a survey of the variety of habitats provided by soil, the variety of soil organisms, the diversity of their life-cycles and phenology, and scale effects, the earthworms are considered as an example. Earthworms are typically stratified vertically into ecological groups, and the niches they occupy can be drastically changed by agricultural practices. The importance of genetic heterogeneity within earthworm species is overlooked, although this can facilitate their spread into new habitats. There is a general reduction of earthworm species in agricultural soils as compared with similar soils under natural vegetation, but there is evidence that at least in some cases the introduction into and management of earthworms in agricultural soils can be beneficial. The requirements for successful introduction programmes and the ideal earthworm spectrum are considered. Opportunities exist for the management of soil organisms for maximum benefit in agro-ecosystems, especially in the tropics.

Introduction

The complexity of relationships between the soil environment and its biota makes it necessary to confine discussion to a limited range of organisms. In the context of this meeting, whose aim is to discuss the relevance of bio-diversity to sustainable agriculture, especially in the tropics, it is appropriate to distinguish from the remainder of the soil biota those few groups, especially earthworms, termites, ants, some insect larvae and a few others of the

The Biodiversity of Microorganisms and Invertebrates: Its Role in Sustainable Agriculture.
Edited by D.L. Hawksworth. © CAB International 1991.

larger soil animals, that are numerous and widespread and have the ability to reshape the soil to suit their requirements. In addition to their influence on organic matter decomposition and nutrient cycling in soils, these animals modify soil structure, porosity, and thus aeration, water infiltration and drainage. Their activities have important consequences for tillage practices, retention and incorporation of plant detritus, the control of surface runoff and erosion. If the best use is to be made of ecological knowledge as a basis for sustainable agriculture, these soil animals have a special significance. Most of this paper concerns earthworms, because they are probably the most widely distributed and best known of these groups.

The variety of habitats provided by soil

Vannier (1973) proposed the term 'porosphere' to describe the environment provided by porous media, that is more or less solid media with internal surfaces, such as soils.

The solid components of soils include mineral particles that range in size from clay ($< 2 \mu m$) particles to large masses of rock, organic materials that include plant litter and logs at the soil surface, dead and living plant roots, dead tissue sloughed from plant roots, microbial and fungal tissue, dead and living bodies of soil invertebrates. Soil organic matter, which provides the basic food supply of all soil organisms, ranges from undecomposed plant tissue to the resistant residues (humic materials) that remain after microbial and faunal decomposition of plant tissue. Most of the soil organic matter is near the surface, with the concentration falling rapidly with increasing depth; some of it is bonded electrostatically to the surfaces of clay particles or incorporated with mineral particles in aggregates.

The spaces between the solid components of soils, making up the porosphere, range in dimensions from large cracks, through macropores that remain where roots have died and decayed or that are the burrows of the larger soil fauna, to minute spaces within soil aggregates. They may be wholly or partially air- or water-filled and the soil water includes ionic and non-ionic solutes, some of which are essential nutrients for the soil biota, some of no significance, while some may be toxic. These spaces, and especially the surfaces that define their limits, provide the habitats of the smaller soil organisms. Their geometry and connectivity define the spatial limits of the niches available to most of the soil biota. Their favourableness as an environment varies in time and space at a variety of scales, determined by variability in litter and soil depth, plant cover, seasonal and diurnal variations in temperature, the proportion of air- to water-filled pore spaces, availability of oxygen, pH, concentration of electrolytes, redox potentials, drainage, chemical composition of the inorganic and organic components of the soil, the nature of the underlying material, whether solid rock, fractured rock or weathered material.

The variety of soil organisms

Representatives of all groups of microorganisms and fungi, green algae and cyanobacteria, and of all but a few exclusively marine phyla of animals, make up the soil biota. An indication of the biomass of soil organisms, divided on the basis of their size into microfauna, mesofauna, macrofauna, and microflora, and comparing the biota of temperate with tropical regions is provided in Table 7.1. This aspect is considered further by Lal (Chapter 8).

The total number of species of all groups of soil-inhabiting organisms has not been and could not be determined for any site. Sampling methods necessarily differ from one group of organisms to another. The results obtained by the different methods are not comparable and for some (perhaps all) groups of organisms are known to underestimate numbers and variety of species. For example, it is generally considered by microbial ecologists that culture methods used for studying soil microorganisms isolate only a small proportion, probably less than 20%, of the species present. Using cloned 16S rcDNA sequences from the cyanobacterial mat at Octopus Spring, Yellowstone National Park, and comparing them with 16S rRNA sequences of microorganisms isolated from the same and similar habitats, Ward *et al.* (1990) demonstrated that although the community of microorganisms in and associated with this mat has been intensively studied by microbiologists, it includes a large number of taxa that are not isolated by the range of culture methods currently available.

Populations of soil organisms vary through the year in response to seasonal and shorter term variations, for example in quality and quantity of

Table 7.1. Representative composition of soil biomass in temperate and tropical zones.

Organisms classified by size	Biomass (kg ha^{-1} live weight)	
	Temperate zone	Tropical zone
Microfauna (< 2 mm)	50	0.5
(protozoa, nematodes)		
Mesofauna (2–10 mm)	20	20
(enchytraeids, microarthropods)		
Macrofauna (> 10 mm)		
Earthworms	900	300
Termites	—	15
Microorganisms		
Bacteria, fungi, and other	20 000	?
microorganisms		

available food, and in abiotic factors, especially temperature and soil moisture. The responses of individual taxa to these variations differ, so that a sampling programme suited to the isolation and quantitative assessment of some taxa cannot be expected to be suited to others.

Diversity of life-cycles and phenology

Most soil organisms are opportunistic in their use of resources. The smaller organisms, such as bacteria, fungi, nematodes, and some other microfauna, can virtually suspend metabolic activity and survive, some of them for many years, as spores, cysts or other resistant structures, in response to lack of suitable food supplies or to adverse physical or chemical conditions in their immediate environment. When suitable food substrates are presented or more favourable environmental conditions are restored they become active and may reproduce in minutes, hours or days.

Some macrofauna, for example, the common pantropical earthworm *Pontoscolex corethrurus* in the seasonally dry soils of northern Queensland, survive drought by leaving cocoons whose hatching is delayed until the onset of rain, or by burrowing deep into the soil, where they seal themselves into a mucus-lined cavity and remain inactive through the dry season. During their period of inactivity the sexual characters regress and the tissue lost is apparently metabolized. Byzova (1977) showed that the cosmopolitan temperate zone earthworm *Aporrectodea caliginosa* quickly uses up all of its glycogen reserves and that there is a reduction of up to 40% in the blood pigment (erythrocruorin), which is apparently also used as an energy source, after 3 months' aestivation. Earthworms that pass through adverse periods in resting stages rapidly return to reproductive activity when soil moisture again becomes favourable.

At the other extreme, some soil-inhabiting insects, such as the larvae of cicadas and melolonthid beetles, have less flexibility and once the life-cycle is initiated, although it may progress slowly when conditions are adverse, must go through to success or failure.

Scale effects

Whittaker (1972) distinguished three basic kinds of diversity:

1. Alpha-diversity, the diversity of species within a community or habitat.
2. Beta-diversity, a measure of the rate and extent of change in species along a gradient from one habitat to others.
3. Gamma-diversity, the richness in species over a range of habitats in a geographical region.

Southwood (1978) recognized gamma-diversity as a consequence of the alpha-diversity of particular habitats together with the extent of the beta-diversity between them.

At the level of alpha-diversity, differences in the dimensions of soil organisms parallel scale differences in the variability of the soil environment. Patterns of environmental and biological diversity that are apparent on a broad scale, as when comparing one sampling site with another, may be seen to repeat at sampling points within sites, and with increasing magnification to continue to repeat, down to levels of observation that are possible only with electron microscopy. Within-site diversity follows a fractal-like model, with compartments that are of similar spatial, physico-chemical and biological complexity, but of diminishing dimensions (Fig. 7.1).

Earthworms as an example

Earthworms are found in all but the driest and the coldest land areas of the earth, and include about 3 000 species. Their biomass is commonly hundreds of kilograms or tonnes per hectare, and their role in promoting soil fertility makes them probably the most important, in terms of their relevance to productivity, of all animal groups that share with mankind the earth's land surface (Bouché, *in* Lee, 1985).

Ecological groups

Earthworm communities are typically stratified vertically into ecologically distinct groups of species that usually do not relate closely to taxonomic categories. Three major groups, with a number of subgroups, were originally recognized in forest ecosystems in New Zealand and in European woodlands (Lee, 1959; Bouché, 1977); these are summarized in Figure 7.2, and are discussed in detail in Lee (1985). Although the three groups are distinguished primarily on the basis of the vertical stratification of their preferred habitats in the soil, species in each of the groups share a variety of morphological, behavioural and demographic characters (Table 7.2).

All three groups are not necessarily to be found in all ecosystems that have earthworms but, using the classification (Fig. 7.2), it is possible to assign the species at many sites to one or other of the groups and consequently to better define the niche and assess the function of each species in the ecosystem.

Niche width and niche overlap

The establishment of agricultural systems involves clearing of the original vegetation, usually some cultivation, followed by monoculture or at least a

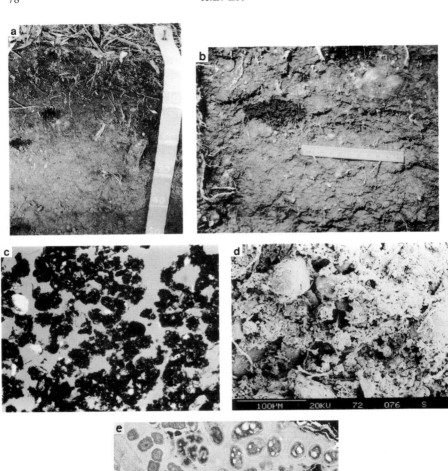

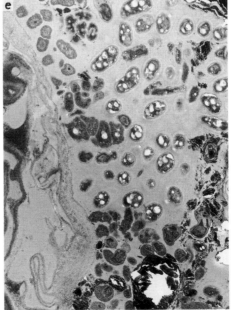

change to ecosystems with less plant species, different from those in the original systems, and a reduction in habitat complexity compared with that prevailing in the original undisturbed ecosystems. These changes result in the destruction or drastic modification of some niches in the soil, and the disappearance of some soil organisms, especially those whose niche widths are narrowly limited. In the case of earthworms and many other soil invertebrates, most or all of the original species disappear and they may or may not be replaced by exotic species that are able to move into empty niches. Most successful among the earthworms that invade newly developed agricultural lands have been a small group of originally European Lumbricidae in temperate regions and a small group of Glossoscolecidae and Megascolecidae, originally from South and Central America, Africa and South-East Asia, in tropical regions. They are opportunistic species, with a capacity to expand or modify their apparent niche width to suit changed circumstances, which makes them most successful colonizers of disturbed and uncrowded habitats – they are typical *r*-selected species (Lee, 1985, 1987). There is much evidence that their dispersal has resulted very largely from accidental transportation with plants spread around the world by humans.

The concepts of niche width and niche overlap (Giller, 1984) are of considerable relevance in monocultural systems of land use. Niche width implies a multidimensional resource utilization spectrum, defined in terms of niche dimensions that are considered to be significant, while niche overlap concerns the tendency for species in a habitat to share parts of each other's fundamental niches, resulting in simultaneous demands upon some resource by two or more species populations, so that some compromise may be necessary for one or more of the species in their use of that resource.

A low level of diversity, accompanied by increases in the variety and width of niches occupied, is apparent in communities of earthworms that have been introduced into newly developed agricultural lands, when compared with the communities that inhabited such lands before they were

---

Figure 7.1. Scale effects in environmental and biological diversity. A fractal-like pattern can be seen with increasing magnification. (a) Soil profile, red brown earth, near Adelaide, South Australia (scale divided into 10-cm intervals). Photo: R.W. Fitzpatrick. (b) A–B horizon transition zone from profile in (a), showing detail of major structural aggregates, mixing of dark organic matter rich A-horizon material (probably earthworm casts) with B-horizon material (scale 10-cm long). Photo: R.W. Fitzpatrick. (c) Detail of mixing of dark organic matter rich material (excreta of microfauna) with paler inorganic material (width of photograph approximately 4 mm). Photo: C.J. Chartres. (d) Scanning electron micrograph, A horizon, showing plant roots and fungal hyphae growing between and around soil micro-aggregates. Photo: R.C. Foster. (e) Transmission electron micrograph, *Trifolium subterraneum* root surface/soil interface. Root surface along left side of frame; clay particles at bottom right; space between root and clay has at least 10 cytologically and morphologically distinct forms of bacteria (frame of photograph $17 \times 8\,\mu m$). Photo: R.C. Foster.

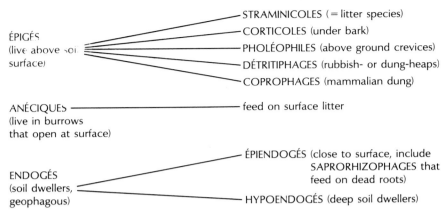

Figure 7.2. Major ecological groups and subgroups of earthworms and their habitats. After Bouché (1977).

developed for agriculture, or with the communities characteristic of agricultural lands in the regions from which the new populations have been introduced (Table 7.3). This would seem to be a consequence of the capricious nature of their transportation to new locations and it favours aggressive colonizers, with a high level of physiological, genetic, behavioural and trophic plasticity.

For example, *Aporrectodea caliginosa* and/or the closely related *A. trapezoides*, which are the most widespread and numerous species in New Zealand pasture soils, are feeders on dead roots, as they are in Europe, where they would be classified as epiendogeic species in the scheme proposed by Bouché. In New Zealand, however, they commonly feed also on dead plant remains

Table 7.2. Characteristics of earthworm species that correlate with the ecological groups recognized by Lee (1959) and Bouché (1977).

Correlates of Major Ecological Groups of Earthworms	
Morphological	**Demographic**
Size	Cocoon production
Pigmentation	Generation time
Muscle development	Length of life
Gut structure	
Behavioural	**Environmental**
Form of burrows	Food supply
Casting behaviour	Predatory pressure
Mobility	Effect of changes
Reaction to desiccation	

Table 7.3. Comparison of numbers of earthworm species in some natural ecosystems and agricultural ecosystems.

Location	Vegetation	Number of species	Earthworm groups
UK	Beech forest	10	Lumbricidae
	Pastures	5–15	Lumbricidae
Poland	Oak–hornbeam forest	8	Lumbricidae
	Pastures	5	Lumbricidae
France	Deciduous forests	9	Lumbricidae
	Pastures	9–14	Lumbricidae
New Zealand	Native forests	1–7	Megascolecidae
	Pastures	1–4	*Lumbricidae
Egypt	Cultivated/pastures	1–2	*Lumbricidae
Venezuela	Tropical rain forest	8	Glossoscolecidae
Australia	Sugarcane	1	*Glossoscolecidae

*Introduced species.

and the dung of herbivores at the soil surface, depositing their casts on the surface, in their burrows and in other spaces below the surface (Lee, 1959). In Bouché's scheme their feeding, burrowing and casting behaviour in New Zealand would make them part endogé, part anécique and part épigé. *Lumbricus terrestris* is rarely found in New Zealand, while in Europe this species is widespread and is frequently the dominant anécique. Some small species, especially *Dendrodrilus rubidus* and *Lumbricus castaneus*, are commonly found in Europe, feeding on the dung of herbivores. Although *D. rubidus* and *L. castaneus* are known from New Zealand, and they feed on the dung of herbivores there, they also are not widely distributed. It seems that in New Zealand the general absence or rarity of species that occupy some niches in Europe has allowed *A. caliginosa* and/or *A. trapezoides* to exploit resources that are not so readily available to these species in Europe.

Recent work in New Zealand (Springett, 1991) indicates that the diversity of lumbricid species increases with increasing time since native vegetation was removed and farms were established. In the north of the North Island, where permanent settlements were first established in New Zealand about 150 years ago, more than 50% of farms were shown by Springett to have three or more species of lumbricids, while in the south of the South Island more than 70% of farms had two species or less. It is particularly interesting that *L. terrestris* is common only in areas in the north of the North Island that were settled in the 19th century and were planted at that time with trees transported in soil from Europe. It may be that the present predominance of a few species in most New Zealand agricultural lands represents only an early stage in a succession that involves increasing diversity and perhaps a reduction in niche width of the earlier established species.

Genetic heterogeneity

Studies of communities, and of the behavioural capabilities of species within them, tend to disregard the possible significance of infraspecific genetic heterogeneity. The niche width of a particular species, its behaviour and its relationships with other species and with abiotic environmental factors, once defined with reference to particular communities in particular ecosystems, are treated as though they are constant, even when the species is in another community of different species composition in a different ecosystem. That this is not always the case can be seen, for example, from the broader niche width achieved in New Zealand by *A. caliginosa* and *A. trapezoides* when compared with these species' niche width in Europe.

Polyploid races are recognized in many species of earthworms and are common among the most widespread of lumbricids (Lee, 1987). Omodeo (1952, 1955) found 13 polyploid species among 29 species of lumbricids. In nearly all polyploid races of earthworms parthenogenesis is obligatory since spermatogenesis is unsuccessful. There is an obvious advantage for obligate parthenogenetic forms, compared with diploid, bisexual forms of the same species, in achieving widespread dispersal after introduction to a new habitat. Further, it was noted by Omodeo (1952) that, within species that include parthenogenetic races, polyploids were more widespread than were sexually reproducing diploids. Jaenike *et al.* (1982) found a number of clones of pentaploid parthenogenetic *D. rubidus* in the north-eastern USA. They noted that pentaploid populations of this species are not known from Europe, which might be taken to indicate that the species is native to North America as well as to Europe. It might as well be taken to indicate that this pentaploid race has an advantage over diploid, bisexual forms of the same species in its dispersal after introduction.

Preferential selection from accidentally introduced and dispersed populations, based on the existence of genetic heterogeneity, may be inferred from the temperature preferenda of common introduced species. For example, the optimum temperature preference of field populations of *Aporrectodea rosea* in Germany was shown by Graff (1953) to be 12°C, while Reinecke (1975) demonstrated that South African representatives of this species had an optimum range of 25.0–26.9°C in laboratory conditions. There is an obvious advantage in having an ability to tolerate higher temperatures for South African compared with European populations.

The technique of DNA fingerprinting is being applied to southern Australian lumbricid populations (G.H. Baker and M. Bull, personal communication), predominantly comprised *A. trapezoides* and *A. rosea*, with the aim of determining whether varieties that are specially adapted to the mediterranean climatic conditions may have been differentiated from the original accidentally introduced populations. The intention is that, if such varieties are shown to exist, they might be spread more widely through southern Australian agricultural lands, to promote pastoral and agricultural production.

Relevance of biodiversity to the stability of populations in agricultural ecosystems

Whether it is necessary or feasible to emulate in an agricultural ecosystem the biodiversity that may be evident in a natural ecosystem is a question that has no answer. For soil organisms, in no case has the full diversity of a soil population been determined, or is it known how optimum biodiversity for an agricultural ecosystem might be defined, or if it were to be defined, how it might be compared with a similarly defined optimum for a non-agricultural ecosystem.

It is reasonable, however, to examine the functional significance of particular groups of organisms as they affect plant production and its sustainability in an agricultural context, to attempt to define the level of diversity that may be expected to optimize the beneficial effects of those groups, and then to work towards the establishment and maintenance of at least that level of diversity for the particular groups.

There is a general reduction in diversity of earthworm species in agricultural land compared with the diversity of species in comparable soils under native vegetation (Table 7.3).

In New Zealand, Stockdill (1966, 1982) introduced *A. caliginosa* to the soils of pastures that previously lacked earthworms and probably had no significant earthworm populations since they were first converted from native tussock grassland to pasture about 6 years previous to the earthworm introduction. Five years after introducing *A. caliginosa*, Stockdill showed that pasture litter, animal dung, surface-applied fertilizers and insecticides had been incorporated into the soil; the organic matter content of the 0–2.5 cm soil layer was much reduced, while that of the 2.5–7.5 cm layer was corresponding increased, because a former surface mat of grass roots had been incorporated into the upper 7.5 cm layer by the feeding, burrowing and casting activities of *A. caliginosa*. Dry matter production from the pasture had increased 19% 2 years after introduction and 28% 10 years after introduction of *A. caliginosa*, while the rate of infiltration of water into the soil increased by close to 100% 10 years after introduction.

In the sugarcane fields of north-eastern Queensland, green harvesting, with associated trash retention and reduced cultivation, have resulted in inputs of up to 15 tonnes ha^{-1} year^{-1} (dry weight) of coarse leaf and stem litter. The pantropical glossoscolecid earthworm *Pontoscolex corethrurus* has been accidentally introduced to the region and is common in the sugar fields.

It feeds near the surface and deposits casts above the soil surface in the litter. Measurements made in a sugarcane field near Innisfail (A.V. Spain and K.E. Lee, personal communication) show that 81% of the litter layer (15 tonnes ha^{-1}) had disappeared 338 days after harvesting and that its disappearance was apparently due to its burial by the casts of *P. corethrurus*, with further decomposition of the buried material presumably associated with soil microorganisms.

Measurements of penetration resistance and the shear strength of soils under rain forest in north-eastern Queensland were compared with measurements where these soils had been converted to sugarcane culture (Spain *et al.*, 1990). Agricultural use, with intense cultivation and the burning of trash after harvesting, resulted in surface compaction, which was associated with traffic by heavy agricultural machinery, and in excessive loss of soil by erosion. Trash retention, associated with green harvesting of the sugarcane and minimum cultivation, resulted in substantial increases in populations of the earthworm *P. corethrurus* and amelioration of the soil structural problems.

The New Zealand and the north-eastern Queensland examples illustrate the benefits to soils and to agricultural production that can be derived from the introduction of one species. In the New Zealand example there is evidence of a broadening of the niche typical of *A. caliginosa* in Europe to take advantage of the absence of other earthworms from the New Zealand pastures.

Management of populations for maximum benefit

There is good evidence that the introduction of earthworms into some agricultural and pastoral soils has resulted in worthwhile improvements in soil fertility, control of erosion losses, plant growth and sustainability of production (Lee, 1985). Existing populations can be manipulated, or new populations can be established, by introducing exotic species. Methods for broad-scale introduction have been developed by Stockdill (1982) in New Zealand, but they are known to be successful for only a few species and in a few localities; further work is needed to develop methods of introduction.

A successful introduction programme depends on:

1. A good knowledge of the physico-chemical and biological characteristics of the soils and of other parameters that will significantly affect the survival, fecundity, potential population density and behaviour of earthworms introduced into the new locality.
2. An informed judgement, based on a knowledge of the requirements for good soil management and factors that control and limit productivity at the new locality, of whether problems have been identified that are amenable to

solution through the activities of earthworms, and whether it may be necessary to modify the soil to provide an environment where earthworms will thrive.

3. An informed judgement, based on a comprehensive knowledge of earthworm ecology, behaviour and physiology, of whether earthworm species already present are capable of providing a high level of benefit, or whether an attempt to introduce new earthworms has a good chance of success.

4. Knowledge of the availability and location of earthworm species that are apparently well-adapted to the new environmental conditions.

5. Understanding of the behaviour of candidate species, especially the depth at which they are active in the soil, their burrowing and casting behaviour, and their food preferences.

6. Establishment of large-scale culture methods for candidate species, or some other method for obtaining large numbers of individuals, and practical methods for their introduction on a scale appropriate to the needs of the new locality and its management.

Given this information, it should be possible to match species with environments and so to select species that are particularly suited for introduction to any particular climatic region or soil management scenario.

An ideal earthworm population might reasonably include:

1. One or more anecic species that feed and deposit their casts at the soil surface, so removing or burying plant litter, and that make vertical or near vertical burrows that facilitate water entry and gas exchange from the soil surface.

2. One or more endogeic species that feed on organic matter already below the soil surface, and that make more horizontal burrows, so facilitating the lateral movement of water and gases within the soil.

3. Among these, or in addition, at least one species that can penetrate compact soil layers.

It should, however, be noted that one species, such as *A. caliginosa* or *P. corethrurus*, may be capable of expanding its 'normal' niche so that it might alone be sufficient to provide the beneficial effects that might otherwise be expected of several species in their original environment. Unfortunately, earthworm ecologists are not well equipped with knowledge on which to base the selection of appropriate species; no more than about six lumbricids and about six tropical species have been yet studied in sufficient detail.

Conclusions

Ecologically based studies of the soil biota and the relationships between soil organisms and soil fertility and plant growth have, up to now, been conducted mainly in temperate regions. The soils of tropical regions are generally more strongly leached and less fertile than those of temperate regions. Their management must be seen against a background of low capital inputs in economies that lack large material resources, coupled with a need to promote and increase sustainable productivity to feed often rapidly increasing human populations.

There is an opportunity, especially in tropical regions but also in other regions, to manage soil organisms for maximum benefit in agro-ecosystems. It requires increased attention to studies of the diversity and functional ecology of the soil biota and it represents the greatest challenge that now faces soil biologists.

References

Bouché, M.B. (1977) Stratégies lombriciennes. *Biological Bulletin, Stockholm* 25, 122–132.

Byzova, J.B. (1977) [Haemoglobin content of *Allolobophora caliginosa* (Sav.) (Lumbricidae, Oligochaeta) during aestivation.] *Doklady Biological Sciences* 236, 763–765.

Giller, P.S. (1984) *Community Structure and the Niche*. Chapman & Hall, London.

Graff, O. (1953) Die Regenwurmer Deutschlands. *Schriften und Forschungen der Landwirtschaft Braunschweig-Volkenrode* 7, 1–81.

Jaenike, J., Ausabel, S. and Grimaldi, D.A. (1982) On the evolution of clonal diversity in parthenogenetic earthworms. *Pedobiologia* 23, 304–310.

Lee, K.E. (1959) The earthworm fauna of New Zealand. *Bulletin, New Zealand Department of Scientific and Industrial Research* 130, 1–486.

Lee, K.E. (1985) *Earthworms: Their Ecology and Relationships with Soils and Land Use*. Academic Press, Sydney.

Lee, K.E. (1987) Peregrine species of earthworms. In: Bonvicini, A.M. and Omodeo, P. (eds), *On Earthworms*. UZI, Modena, pp. 315–327.

Omodeo, P. (1952) Cariologia dei Lumbricidae. *Caryologia* 4, 173–275.

Omodeo, P. (1955) Cariologia dei Lumbricidae. II. *Caryologia* 8, 135–178.

Reinecke, A.J. (1975) The influence of acclimation and soil moisture on the temperature preference of *Eisenia rosea* (Lumbricidae). In: Vanek, J. (ed.), *Progress in Soil Zoology*. [Proceedings of the 5th International Colloqium on Soil Zoology, Prague, 1975.] Junk, The Hague, pp. 341–349.

Southwood, T.R.E. (1978) *Ecological Methods*. Chapman & Hall, London.

Spain, A.V., Prove, B.G., Hodgen, M.J. and Lee, K.E. (1990) Seasonal variation in penetration resistance and shear strength of three rainforest soils from northeastern Queensland. *Geoderma* 47, 79–92.

Springett, J.A. (1991) Distribution of lumbricid earthworms in New Zealand. *Soil Biology and Biochemistry*, in press.

Stockdill, S.M.J. (1966) The effect of earthworms on pastures. *Proceedings of the New Zealand Ecological Society* 13, 68–75.

Stockdill, S.M.J. (1982) Effects of introduced earthworms on the productivity of New Zealand pastures. *Pedobiologia* 24, 29–35.

Vannier, G. (1973) Originalité des conditions de vie dans le sol due à la présence de l'eau, importance thermodynamique et biologique de la porosphère. *Annales de la Société de Zoologie Belgique* 103, 157–167.

Ward, D.M., Weller, R. and Bateson, M.M. (1990) 16S rRNA sequences reveal numerous uncultured microorganisms in a natural community. *Nature* 345, 63–65.

Whittaker, R.H. (1972) Evolution and measurement of species diversity. *Taxon* 21, 213–251.

Eight

Soil Conservation and Biodiversity

R. Lal, *Department of Agronomy, The Ohio State University, Columbus, Ohio 43210, USA*

ABSTRACT: Accelerated soil erosion is a symptom of land misuse. The latter, due to drastic simplification of an ecosystem and reduction in activity and species diversity of the soil biota, leads to degradation of soil structure and increase in runoff and erosion. Erosion risks can be reduced through the ameliorative effects of soil biota on water transmission properties. Soil biota influence soil properties through formation of stable aggregates, development of organo-mineral complexes, and bonding through fungal hyphae and polysaccharides. Improvements in properties are also brought about by mixing soil with organic residues and by turnover through burrowing and the formation of subterranean chambers and feeding galleries. These activities, by improving macroporosity and continuity of pores from surface to the subsoil, increase the infiltration rate and reduce runoff rate and amount. Agricultural practices that enhance the activity and species diversity of the soil biota may reduce the risks of runoff and soil erosion. Such practices include afforestation and establishment of pasture, land development through manual rather than mechanical means, mulch farming and conservation tillage, mixed cropping and agroforestry, frequent use of legume/grass cover crops in the rotation, and discriminate and judicious use of agrochemicals. Biological systems that enhance the activity of the soil biota are usually more effective in mitigating soil erosion than engineering practices of water management.

Introduction

Accelerated soil erosion is a global problem, and a major economic and

The Biodiversity of Microorganisms and Invertebrates: Its Role in Sustainable Agriculture.
Edited by D.L. Hawksworth. © CAB International 1991.

environmental issue of modern times. The problem is particularly severe in ecologically sensitive regions where steep slopes and marginal lands are being cultivated intensively. Examples of the most severely affected regions include the Himalayan–Tibetan ecosystem (Lal, 1990) and the Yellow River Valley of China (Robinson, 1981), marginal sloping lands in India (Narayan *et al.*, 1985), Nepal (Hurni, 1985), north-eastern Thailand (Hurni and Nuntapong, 1983), highlands of Ethiopia (Hurni and Messerli, 1981; Hurni, 1983), the Caribbean (Ahmad, 1977), and the Pacific region (Clarke and Morrison, 1987), Australia (Messer, 1987), Europe (Chisci, 1986; Eppink, 1986; Schwertmann, 1986), North America (Batie, 1983; Gardner, 1985; Camboni and Napier, 1986; Phipps, 1988) and Africa (Lal, 1990).

Severe erosion has both edaphological and ecological implications. The edaphological impact is due to its on-site effects causing losses in crop yields and reductions in profit margin. The environmental impact is due to the eutrophication of natural waters, transport of dissolved and suspended solids, and the siltation of waterways and reservoirs. Gully erosion has scarred the landscape in highland and in semi-arid regions. Globally, productivity of some 630 million hectares of arable land has been severely jeopardized by past erosion. Furthermore, a high proportion of newly cleared land in the tropical rain forest ecosystems is subject to the degrading effects of accelerated erosion. Soil erosion has attained the level of a global crisis, and is a cancer of the land.

Soil structure and biodiversity of soil fauna

Soil conservation implies mitigating the erosive effects of impacting raindrops, running water or wind. Diminuition of the erosive impact can be achieved by: (i) dissipating the kinetic energy of agents of erosion; and (ii) by increasing the strength of structural aggregates. Diversity of the soil biota minimizes the risks of soil erosion through both of these mechanisms. The safe disposal of excessive overland flow through engineering technologies is another important strategy for reducing erosion risks. The effectiveness of engineering technology by itself, however, is questionable. The problem of soil erosion has not been solved despite the billions of dollars invested in erosion control measures throughout the world. At least 50% of the soil conservation projects in Africa have been considered unsatisfactory because of their ineffectiveness in conserving soil and water resources in sustainable land use. The basic problem is to develop systems that enhance soil structure, improve the infiltration rate, decrease the runoff rate and velocity, and enable the structural units of the soil to resist the erosive forces of rainfall and runoff. Diversity of the soil biota has a potential to impart these favourable characteristics to soil.

Diversity in soil biota

Soil is a habitat for vast numbers of diverse species, some of which are yet to be identified. In fact, soil is a living entity with a diverse fauna and flora. When devoid of its biota, the uppermost layer of earth ceases to be a soil. Any definition of soil is, therefore, incomplete unless it includes reference to the activity and species diversity of soil organisms (Fig. 8.1).

In relation to soil conservation, the soil biota can be classified into the following categories (Hole, 1981; Lee, Chapter 7):

1. Microflora comprises bacteria, actinomycetes, fungi, and algae. Their biomass ranges from 1 to $100 \, g \, m^{-2}$ and their number varies up to $1\,000\,000$ million m^{-2}.
2. Microfauna comprises mostly protozoa with a biomass range of 1.5–$6 \, g \, m^{-2}$. Their size is greater than 0.2 mm and their number ranges from 1 million to $1\,000\,000$ million m^{-2}.
3. Mesofauna includes acari, collembola, nematodes, etc. Their number ranges from a few hundred to several million m^{-2}, biomass from 0.01 to $10 \, g \, m^{-2}$ and size from 0.2 to 10 mm.
4. Macrofauna primarily consist of insects. Their biomass ranges from 0.1 to $2.5 \, g \, m^{-2}$, number from a few hundred to several thousands, and size from 10 to 20 mm.
5. Megafauna consists of earthworms with a range in number from a few to several hundreds m^{-2} biomass weight from 10 to $40 \, g \, m^{-2}$, and size generally exceeding 20 mm.

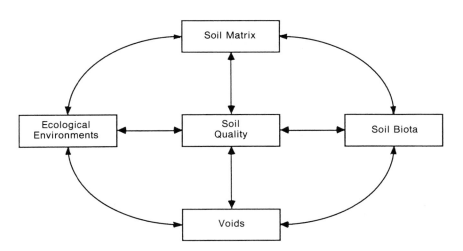

Figure 8.1. Soil in a dynamic equilibrium with its different phases and ecological environment.

The macro- and megafauna are classified as 'endopedonic' animals because they live inside the soil. The animals that live above the soil or within the leaf litter are called 'exopedonic'. Activities of the soil biota that affect soil erosion and soil structure are the ingestion and grinding of soil particles, soil mixing and turnover, casting and mounding or nest-building, and burrowing or tunnelling. These activities lead to humification, synthesis of organo-mineral complexes, formation of macropores, and the development of stable aggregates.

Soil structure and organo-mineral complexes

Soil structure and soil biota are interdependent, that is soil structure affects the soil biota and *vice versa*. Soil structure affects the biota through its

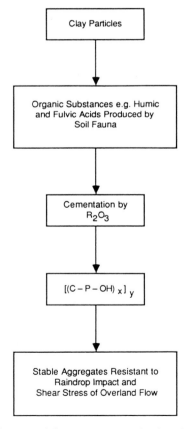

Figure 8.2. A schematic diagram of the processes involved in the formation of structural aggregates through by-products of the biological activity of the soil fauna.

influence on aeration, drainage, water retention, and living space. Biota affects the soil structure through alterations in mechanical and biochemical processes.

Activity of the soil biota produces organic polymers that bind the clay particles into domains, and then domains into micro-aggregates (Emerson, 1959; Williams *et al.*, 1967; Tisdall and Oades, 1982). There are several mechanisms of bonding between humic substances and clay mineral surfaces. Some of these mechanisms include ion–dipole, cation bridging, water bridging, anion exchange, ligand exchange, hydrogen bonding, van der Waals' force, and entrophy effects (Theng, 1979; Stevenson, 1982). The anionic character of organic acids is pH dependent and facilitates their absorption on negatively charged clay particles through H-bonding and van der Waals' forces. Organo-mineral compounds exist in soil as polymeric complexes of humic and fulvic acids bound together by polyvalent cations (Ca^{2+}, Fe^{3+}, and Al^{3+}). These complexes are responsible for the development of structural units or aggregates. The bonding is facilitated by the presence of polyvalent cations on the exchange complex, because cations neutralize negative charges on the clay as well as the acidic group of the organic molecules to form a salt bridge:

clay–polyvalent cation–humus

The humic fraction is bound through cation and anion exchange and other mechanisms. A schematic diagram of the processes involved in the formation of aggregates by organo-mineral complexes is shown in Fig. 8.2. These initial bondings are further strengthened by the activity of soil organisms.

Fungal polysaccharides and other byproducts of microorganisms

The activity of a diverse range of organisms (including bacteria, filamentous fungi, yeasts, and algae) increases soil aggregation. The aggregating ability of different microorganisms, however, differs widely among species and depends on the food substrate (Aspiras *et al.*, 1971).

Fungal hyphae and polysaccharides of microbial origin play an important role in soil aggregation (Martin, 1945; MacCalla, 1946; Harris *et al.*, 1960), and aggregates stabilized by fungal hyphae can be strong enough to resist even sonification treatment (Aspiras *et al.*, 1971), because of some of the mechanical and biochemical mechanisms involved (Dorioz and Robert, 1982). Mechanical forces created by fungal hyphae may be involved in the reorganization of mineral particles into aggregates. Mechanical forces may lead to the initiation of aggregation. Subsequently, aggregates are stabilized by physico-chemical processes. Chenu (1989) confirmed that there is a strong water-stabilizing effect of fungal polysaccharides related to the formation of a stable organo-mineral network. Aggregates may be stabilized through the cementation of extracellular polysaccharides.

Table 8.1. Mean diameter of aggregates of earthworm cast and the adjacent surface soil in no-till systems of corn production (Lal and Akinremi, 1983).

Fertilizer	Mulch	Mean weight diameter (mm)	
		Cast	Soil
Without	Without	6.83	3.16
Without	With	6.72	3.30
With	Without	6.27	1.88
With	With	6.95	2.48
LSD (0.05)		———— 0.36 ————	

The fungi most effective in binding soil particles are rapidly growing species that produce woolly mycelia, such as species of *Absidia*, *Mucor*, *Rhizopus*, *Chaetomium*, *Fusarium*, and *Aspergillus* (Harris *et al.*, 1966). Important aggregate-stabilizing polysaccharides, which are microbially synthesized are periodate-oxidizable polysaccharides (Greenland *et al.*, 1962; Harris *et al.*, 1966). Microbial activity, under favourable conditions, can also stabilize coarse-textured sand (Foster and Nicholson, 1981*a*, *b*).

Earthworms

Earthworms play a very important positive role in improving soil structure; they affect it through burrowing, mixing, and casting. Earthworm casts are very stable against raindrop impact and have a higher mean weight diameter of aggregates than the soil (Table 8.1). Experiments conducted by Temirov and Valiakhmedov (1988) in the USSR also showed that percentage aggregation increased by a factor of 8–9 by inoculation with earthworms (Table 8.2). Worms create tunnels that are continuous and serve as conduits for the rapid transmission of water from surface through the subsoil (Ehlers, 1975; Kemper *et al.*, 1987; Zachmann *et al.*, 1987). Earthworms can burrow tunnels even in media at a high soil strength. Dexter (1978) observed that tunnelling

Table 8.2. Influence of earthworms on per cent aggregation (Temirov and Valiakhmedov, 1981).

Treatment	Relative aggregation (> 0.25 mm) after inoculation	
	1 Month	3 Months
Initial soil	1.00	1.00
Control without earthworms	2.39	3.12
15 earthworms	3.39	7.66
30 earthworms	6.59	8.77

Table 8.3. Infiltration of water into soil containing clover hay and three levels of worm inoculation (*Eisenia foetida*) (Abbott and Parker, 1981).

Clover treatment	Number of worms added					
	0		2		4	
	Nil	Mixed	Nil	Mixed	Nil	Mixed
Mean time (seconds) for 2 cm of water to infiltrate	184	284	177	202	164	178
	ab	a	ab	ab	ab	ab

Figures with the same letter(s) beneath are statistically similar.

by earthworms was independent of soil strength over the range of micro-penetrometer resistance from 0.3 to 3 MPa. Earthworms make tunnels by ingesting soil particles from ahead of them. Therefore, soils with a higher earthworm activity have a higher infiltration rate than those without it. Favourable infiltration rates in worm-infested soils are due to two factors: (i) earthworm activity reduces the probability of the formation of crust and surface seal; and (ii) the macropores thus created conduct water and improve infiltration. Abbott and Parker (1981) demonstrated that some species that move to the surface to feed on crop residues create vertical tunnels and increase infiltration (Table 8.3). Worm tunnels also promote root growth. The latter extend and stabilize these tunnels, especially when the soil is not disturbed.

Some earthworm species, however, have a negative impact on soil structure. Rather than a granular, friable or mull type of soil structure, Thorp (1949) observed that earthworm activity transformed the friable structure of a chernozemic clay soil into a sticky unmanageable mass. In Alberta, Shaw and Pawluk (1986) also observed that some species of earthworms have a negative influence on soil structure. Furthermore, the nature of the effect (positive or negative) also depends on soil type.

Termites

As in the case of earthworms, termites affect the soil structure through the construction of mounds, feeding galleries, and tunnels. The burrowing activity is specially significant in mound-building species. The subterranean galleries may radiate 40–50 m from the mound.

The structural effects of termites may vary depending on the location with respect to the mound. The mound itself is densely packed, has low porosity and a massive structure. The massive structure is created by a cementing together of the soil particles by saliva and other body wastes. The mound surface has a low permeability to water and can generate a large volume of runoff (Omo Malaka, 1977). Lal (1987) observed that accumulative infiltration at 120 min was 8.5 cm on mound-crust compared with 36.5 cm on an adjoining normal soil.

In contrast to the mound, termite-infested soil usually contains feeding galleries and tunnels, and has favourable aggregation properties. In Texas, Spears *et al.* (1975) observed that the percentage of aggregates less than 0.5 mm was greater in termite-infested soil. However, these aggregates were easily dispersed in water. Furthermore, the percentage of aggregations exceeding 2 mm was less in termite-infested soil. In addition, non-capillary pore space was significantly greater in termite-infested than in termite-free soil in summer and winter, but not in spring. The effect of feeding galleries on infiltration rate is not widely studied. The infiltration rate may be improved only if the tunnels and galleries are connected to the surface and to the source of water supply.

Soil biota and erosion

The soil biota can affect erosion through its influence on decomposition of plant and animal remains, soil turnover, loosening, aggregation, and water transmission and retention. Erosion-preventive effects of the soil biota include improvements in aggregation, prevention of surface crust formation, and increase in infiltration capacity. In contrast, the soil biota may also accelerate erosion. For example, soil freshly turned-over by animal activity is easily displaced downslope by raindrop impact or overland flow. In Kenya, Wielemaker (1984) estimated soil turnover by mole activity as equivalent to 50 tonnes ha^{-1} year^{-1}. He observed that rodent activity alone in an uncultivated soil resulted in an annual displacement of a soil layer with an average thickness of 0.04 mm through a distance of 7 cm. In Brazil, Weber (1966) and Holldobler (1984) estimated that ants turnover some 40 tonnes of subsoil to the surface each year.

Earthworms

Earthworm species with a negative impact on soil structure (Shaw and Pawluk, 1986) enhance slaking and accelerate runoff and erosion. Generally, soils with low or no earthworm activity are more prone to erosion than those with high activity. Sharpley *et al.* (1979) reported that chemical elimination of earthworms doubled the amount of annual runoff from a 13° slope. Furthermore earthworms are washed away in water runoff and eroded sediment (Altavinyte *et al.*, 1974), conversely, eroded soils have lower populations and species diversity of earthworms than uneroded soils.

There are few experiments, if any, that have established a direct cause–effect relationship between activity of the soil biota and erosion. In New

Table 8.4. Runoff and soil erosion from pasture with and without treatment with carbaryl (Sharpley *et al.*, 1979).

Parameter (kg ha^{-1}*)	Carbaryl treatment	
	With	Without
Runoff (cm)	32.1	16.5
Sediment	2.90	11.20
Dissolved inorganic P	1.18	0.31
Total dissolved P	1.49	0.35
Particulate P	0.31	0.56
Total P	1.80	0.91
NH^+-N	9.53	1.63
NO_3-N	4.25	0.52
Total N	17.29	4.73

*Unless otherwise indicated.

Zealand, Sharpley *et al.* (1979) evaluated runoff, sediment discharge and transport of water soluble chemicals from pasture plots with and without an application of carbaryl (a broad-spectrum carbomate biocide). Treatment of soil with carbaryl resulted in a two-fold increase in the volume of surface runoff. Elimination of earthworms decreased the infiltration rate from an initial value of 0.43 to 0.13 mm min^{-1}, whereas the infiltration rate was maintained in untreated plots. Conversely, however, there was a greater transport of sediment from plots with than without earthworms (Table 8.4).

Earthworm casts are rich in water soluble plant nutrients (De Vleeschauwer and Lal, 1981). This implies that runoff from a soil with a high density of earthworm casts may contain more concentrations of nutrients, leading to pollution and eutrophication of water, than those with a low density of surface casts. The data in Table 8.4 show a higher loss in runoff of total phosphorus and dissolved inorganic phosphorus, NH^+-N, NO_3-N, and total nitrogen in plots without earthworms than with earthworms. However, there was substantially more loss of particulate phosphorus from plots with, than without, earthworms. The loss of particulate phosphorus was associated with the greater transport of sediments from worm casts.

Termites

In general, soil in the vicinity of termite mounds is very porous and has a stable structure with a favourable infiltration rate. Such a soil is not prone to erosion even on steep slopes. However, termite activity may accelerate soil erosion through denuding the biomass and vegetation cover, especially in arid and semi-arid regions. In Australia, Watson and Gray (1990) reported that termites were mainly responsible for grass removal. Sheet erosion observed in denuded areas was thus due to the lack of protective vegetation cover.

Crusted surfaces of termitaria generate a significant quantity of surface runoff. Meyer (1960) estimated that as much as 30% of the land surface is covered by active and abandoned termitaria in some regions of tropical Africa. In Burkina Faso, Roose (1976) estimated soil erosion of 600 kg ha^{-1} year^{-1} from abandoned mounds and 800 kg ha^{-1} year^{-1} from active/occupied mounds. Roose estimated the net erosion rate as 0.022 mm year^{-1}. In Uganda, Pomeroy (1976) estimated that one-tenth of the above ground volume of the mound is eroded annually. The annual rate of soil erosion from termite mounds was estimated to be 0.0112–0.10 mm for *Pseudocanthotermes* species, 0.022 mm for *Macrotermes* species, 0.039–0.115 mm for *Bellicosus* species, and 0.026–0.637 mm for *M. subhyalinus*. In Kenya, Wielemaker (1984) estimated soil erosion from termite mounds to be as much as 1163 kg ha^{-1} year^{-1}.

Most of the soil eroded from termite mounds is deposited in their vicinity. Soil in the vicinity of an active mound, therefore, is continuously being enriched by the addition of plant nutrients brought up from the subsoil. It is a natural nutrient recycling mechanism. In Kenya, Wielemaker (1984) estimated the rate of addition of calcium to the soil by erosion from termite mounds as 3–5 kg ha^{-1} year^{-1}.

Conserving soil through biodiversity

The evaluation of the literature presented above indicates that the soil biota has favourable effects on soil structure, macroporosity, and infiltration capacity. Although the transport of sediment and surface runoff may increase due to soil turnover, the mounding or casting activity of soil biota, the overall effect in terms of enhancement of soil structure, increase in infiltration rate, and improvement in macroporosity are positive. This means that an increase in the activity and species diversity of the soil biota can reduce risks of runoff and accelerated erosion, and support a sustainable use of soil and water resources. It is, therefore, desirable to promote agricultural practices that would enhance the activity and species diversity of the soil biota.

Natural ecosystems are known to support a high population of soil macro-invertebrates, and other soil biota (Lee and Wood, 1971; Lee, 1985; Lavelle, 1986). Simple ecosystems, such as arable land-use and man-made ecosystems, usually have a lesser population density of soil biota. Furthermore, the use of agricultural chemicals suppresses the activity, species diversity, and population of the soil biota. Change in land-use generally results in drastic and rapid alterations of vegetation, disturbs and even destroys the habitat, and decreases food availability and diversity. The conversion of forest or range-land to an arable land use alters the ecological equilibrium through drastic

Table 8.5. Effects of land use on soil biota of an ultisol in Yurimaguas, Peru (Lavelle and Pashanasi, 1989).

Land use	Number of taxonomic units	Population density (m^{-3}) (mean ± SE)	Biomass $(g\,m^{-2})$ (mean ± SE)
Forestry			
Primary	41	4 304 ± 933	53.9 ± 8.54
Secondary (15 years)	27	4 099 ± 1 828	24.1 ± 5.85
Crops			
High input maize	20	730 ± 221	3.1 ± 1.04
Low input rice	24	3 683 ± 1 059	8.5 ± 2.75
Traditional cassava	18	1 197 ± 283	7.6 ± 3.26
Pastures			
Brachiaria + *Desmodium*	27	922 ± 81	159.2 ± 16.4
Traditional (moist)	23	1 768 ± 308	121.1 ± 20.8
Traditional (dry)	20	2 367 ± 453	82.3 ± 15.5
Fallows			
Centrosema (6-month)	22	1 546 ± 253	111.9 ± 13.2
Kudzu	23	2 214 ± 969	15.5 ± 6.2
Peach palm + Kudzu	32	1 858 ± 380	93.9 ± 25.2

changes in the water and energy balance, causes high variations in soil temperature and moisture regimes, disrupts cycles of major plant nutrients, and changes the quality and quantity of biomass return to the soil (Lal, 1987).

At Yurimaguas, Peru, Lavelle and Pashanasi (1989) observed that primary rain forest had a rather diverse and abundant soil fauna with a population density of 6303 individuals m^{-2} and a biomass of 53.9 g. In high-input cultivated plots, biomass was reduced to 6% of that in the primary forest site, and the population density to 17%. Furthermore, half of the taxonomic units had disappeared. Traditional and low input systems had similar but less dramatic effects. Even 6-month fallowing with Kudzu reversed these trends leading to a faunal biomass five times more than that observed in high-input cropping (Table 8.5). The data showed drastic detrimental effects of cropping on soil macrofauna especially on earthworms. In contrast, pastures and legume fallows encouraged the recovery of the faunal communities.

Technological options for the sustainable management of soil and water resources are those that maintain or enhance high populations and taxonomic diversity of the soil biota. Some of these practices include manual rather than mechanical methods of deforestation (Lal and Cummings, 1979), sowing cover crops immediately after deforestation (Hulugalle *et al.*, 1984), use of crop residue mulch (Lal *et al.*, 1980), mixed cropping (Okigbo and Greenland, 1976), agroforestry and alley cropping (Kang *et al.*, 1981), and conservation tillage (Lal, 1989a). The activity and taxonomic diversity of earthworms and other soil biota is generally more in no-till than in plough-based farming (Edwards, 1975; Lal, 1976).

There is a need to adopt a systems approach to soil conservation. Agricultural practices that enhance the soil biota should be integrated within routine farm operations so that soil conservation is built into the farming systems and is not a separate endeavour. If that were the case, soil erosion would cease to be a serious global problem.

References

Abbott, I. and Parker, C.A. (1981) Interactions between earthworms and their soil environment. *Soil Biology and Biochemistry* 13, 191–197.

Ahmad, N. (1977) Erosion hazard and farming systems in the Carribbean countries. In: Greenland, D.J. and Lal, R. (eds), *Soil Conservation and Management in the Humid Tropics*. John Wiley & Sons, Chichester, pp. 241–250.

Altavinyte, O., Kuginyle, Z. and Pideckis, S. (1974) Erosion affects on soil fauna under different crops. *Pedobiologia* 14, 35–40.

Aspiras, R.B., Allen, O.N., Harris, R.F. and Charles, G. (1971) The role of micro-organisms in the stabilization of soil aggregates. *Soil Biology and Biochemistry* 3, 347–353.

Batie, S.S. (1983) *Soil Erosion: Crises in America's Croplands?* Resources for the Future Press, Washington, DC.

Camboni, S.M. and Napier, T.L. (1986) Five decades of soil erosion: problems and potentials. [Paper presented at the National Rural Sociological Society meeting, Salt Lake City, Utah.]

Chenu, C. (1989) Influence of fungal polysaccharide, scleroglucan on clay microstructures. *Soil Biology and Biochemistry* 21, 299–305.

Chisci, G. (1986) Influence of change in land use management on the acceleration of land degradation phenomena in the Apennines hilly areas. In: Chisci, G. and Morgan, R.P.C. (eds), *Soil Erosion in the European Community*. A.A. Balkema Publishers, Rotterdam, pp. 3–16.

Clarke, W. and Morrison, J. (1987) Lane mismanagement and the development imperative in Fiji. In: Blaikie, P. and Brookfield, H. (eds), *Land Degradation and Society*. Methuen, London, pp. 176–185.

de Vleeschauwer, D. and Lal, R. (1981) Properties of worm casts in some tropical soils. *Soil Science* 132, 175–181.

Dexter, A.R. (1978) Tunnelling in soil by earthworms. *Soil Biology and Biochemistry* 10, 447–449.

Dorioz, T.M. and Robert, M. (1982) Étude experimentale de l'interaction entre champignons et axiles: consequences sur la microstructure de sols. *Comptes rendus de l'Academic des Sciences, Paris* 295, 511–516.

Edwards, C.A. (1975) Effects of direct drilling on to soil fauna. *Outlook on Agriculture* 8, 243–244.

Ehlers, W. (1975) Observations on earthworm channels and infiltration on tilled and untilled soil. *Soil Science* 119, 242–245.

Emerson, W.W. (1959) The structure of soil crumbs. *Trends in Soil Science* 10, 233–244.

Eppink, L.A.A.J. (1986) Water erosion in the Netherlands: damage and farmer's attitude. In: G. Chisci and Morgan, P.P.C. (eds), *Soil Erosion in the European Community*. A.A. Balkema Publishers, Rotterdam, pp. 173–182.

Foster, S.M. and Nicolson, T.H. (1981a) Microbial aggregation of sand in a maritime dune succession. *Soil Biology and Biochemistry* 13, 205–208.

Foster, S.M. and Nicolson, T.H. (1981b) Aggregation of sand from a maritime embryo sand dune by micro-organisms and higher plants. *Soil Biology and Biochemistry* 13, 199–203.

Gardner, B.D. (1985) Government and conservation: A case of good intention but misplaced incentives. In: *Soil Conservation: What Should be the Role of Government*. Indiana Cooperative Extension Service, Purdue University, West Lafayette, Indiana.

Greenland, D.J., Lindstom, G.R. and Quirk, J.P. (1962) Organic materials which stabilize natural soil aggregates. *Proceeding of the American Soil Science Society* 26, 366–371.

Harris, R.F., Chester, G. and Allen, O.N. (1966) Dynamics of soil aggregation. *Advances in Agronomy* 18, 107–169.

Hole, F.D. (1981) Effects of animals on soil. *Geoderma* 25, 75–112.

Holldobler, B. (1984) The wonderfully diverse ways of ants. *National Geographic* 165, 779–813.

Hulugalle, N.R., Lal, R. and ter Kuile, C.H.H. (1984) Soil physical changes and crop root growth following different methods of land clearing in Western Nigeria. *Soil Science* 138, 172–179.

Hurni, H. (1983) Soil erosion and soil formation in agricultural eco-systems Ethiopia and Northern Thailand. *Mountain Research and Development* 3(2), 131–142.

Hurni, H. (1985) An ecosystem approach to soil conservation. In: El-Swaigy, S.A., Moldenhauer, W.C. and Lo, A. (eds), *Soil Erosion and Conservation*. Soil Conservation Society of America Press, Ankeny, Iowa, pp. 759–771.

Hurni, H. and Messerli, B. (1981) Mountain research for conservation and development in Simen, Ethiopia. *Mountain Research and Development* 1(1), 49–54.

Hurni, H. and Nuntapong, S. (1983) Agro-forestry improvements for shifting cultivation systems soil conservation research in Northern Thailand. *Mountain Research and Development* 3(4), 345.

Kemper, W.D., Trout, T.J., Segeren, A. and Bullock, M. (1987) Worms and water. Journal of Soil and Water Conservation 63, 601–606.

Lal, R. (1976) No-tillage effects on soil properties under different crops in western Nigeria. *Soil Science Society of America Journal* 40, 762–768.

Lal, R. (1987) *Tropical Ecology and Physical Edaphology*. J. Wiley and Sons, Chichester.

Lal, R. (1989a) Conservation tillage for sustainable agriculture. *Advances in Agronomy* 62, 85–197.

Lal, R. (1989b) Soil erosion and land degradation: the global risk. *Advances in Soil Science* 11, 129–172.

Lal, R. (1990) *Soil Erosion in the Tropics*. McGraw-Hill, New York.

Lal, R. and Akinremi, O.O. (1983) Physical properties of earthworm casts and surface soil as influenced by management. *Soil Science* 135, 116–122.

Lal, R. and Cummings, D.J. (1979) Changes in soil and microclimate after clearing a tropical forest. *Field Crops Research* 2, 91–107.

Lal, R., de Vleeschauwer, D. and Nganje, R.M. (1980) Changes in properties of a newly cleared tropical alfisol as affected by mulching. *Soil Science Society of America Journal* 66, 827–832.

Lavelle, P. (1986) The soil system in the humid tropics. *Biology International* 9, 2–17.

Lavelle, P. and Pashanasi, B. (1989) Soil macrofauna and land management in Peruvian Amazonia. *Pedobiologia* 33, 283–291.

Lee, K.E. (1985) *Earthworms: Their Ecology and Relationships with Soil and Landuse*. Academic Press, London.

Lee, K.E. and Wood, T.G. (1971) *Termites and Soils*. Academic Press, London.

MacCalla, T.M. (1946) Influence of some microbial groups on stabilizing soil structure against falling water drops. *Soil Science Society of America, Proceedings* 11, 260–263.

Martin, J.P. (1945) Micro-organisms and soil aggregation. *Soil Science* 59, 163–174.

Messer, J. (1987) The sociology and the politics of land degradation in Australia. In: P. Blaikie and Brookfield, H. (eds), *Land Degradation and Society*. Methuen Press, London, pp. 32–238.

Meyer, J.A. (1960) Resultants agronomiques d'un essaie de nivellement de termitieres réalise dans la cuvette centrale Congolaise. *Bulletin Agricole du Congo et Belge* 51, 1047–1057.

Narayana, V., Dhruva, V. and Sastry, G. (1985) Soil conservation in India. In: El-Swaify, S.A., Moldenhauer, W.C. and Lo, A. (eds), *Soil Erosion and Conservation*. Conservation Society of America Press, Ankeny, Iowa, pp. 3–9.

Okigbo, B.N. and Greenland, D.J. (1976) Intercropping systems in tropical Africa. In: Papendick, R.I., Sanchez, P.A. and Triplett, G.B. (eds), *Multiple Cropping*. ASA, Madison, USA.

Phipps, T. (1988) The conservation reserve and water quality: Results from a simulation study. [Lecture presented at a symposium focussed on the Conservation Title of the Food Security Act of 1985 sponsored by NCR-111, Washington DC.]

Pomeroy, D.E. (1976) Some effects of mound-building termites on soils in Uganda. *Journal of Soil Science* 27, 377–384.

Robinson, A.R. (1981) Erosion and sediment control in China's Yellow River Basin. *Journal of Soil and Water Conservation* 36(3), 125–127.

Roose, E.J. (1976) Contribution à l'ètude de l'influence de le mesofaune sur la pèdogènèse actvelle en milieu tropical. *Rapport ORSTOM, Centre à Apiopoidoumè Ivory Coast*.

Schwertmann, U. (1986) Soil erosion: extent, prediction and protection in Bavaria. In: Chisci, G. and Morgan, R.P.C. (eds), *Soil Erosion in the European Community*. A.A. Balkema Publishers, Rotterdam, pp. 185–200.

Sharpley, A.N., Syers, J.K. and Springett, J.A. (1979) Effect of surface-casting earthworms on the transport of phosphorus and nitrogen in surface runoff from pasture. *Soil Biology and Biochemistry* 11, 459–462.

Shaw, C. and Pawluk, S. (1986) The development of soil structure by *Octolasion turtaeum*, *Aporrectodea tyrgide* and *Lumbricus terrestris* in parent materials belonging to different classes. *Pedobiologia* 29, 327–339.

Spears, B.M., Ueckert, D.N. and Whigham, T.L. (1975) Desert termite control in a short grass prairie: effect on soil physical properties. *Environmental Entomology* 4, 889–904.

Stevenson, F.J. (1982) *Humus Chemistry: Genesis, Composition, Reaction.* J. Wiley & Sons, New York.

Temirov, T. and Valiakhmedov, B. (1988) Influence of earthworms on fertility of high altitude desert soil in Tajikstan. *Pedobiologia* 32, 293–300.

Theng, B.K.G. (1979) *Formation and Properties of Clay Polymer Complexes.* Elsevier, Amsterdam.

Tisdall, J.M. and Oades, J.M. (1982) Organic matter and water stable aggregates in soils. *Journal of Soil Science* 33, 161–163.

Thorp, J. (1949) Effects of certain animals that live in the soil. *Science Monthly* 68, 180–191.

Watson, J.A.L. and Gray, F.J. (1970) Role of grass eating termites in the degradation of a mulga ecosystem. *Search* 1, 43–46.

Weber, N.A. (1966) Fungus growing ants. *Science* 153, 587–604.

Wielemaker, W.G. (1984) *Soil Formation by Termites: A Study in the Kisii Area, Kenya.* Agricultural University, Wageningen.

Williams, B.G., Greenland, D.J. and Quirk, J.P. (1967) The effect of polyvinyl alcohol on the nitrogen surface area and pore structure of soils. *Australian Journal of Soil Research* 5, 77–83.

Zachmann, J.E., Linden, D.R. and Clapp, C.E. (1987) Macroporous infiltration and redistribution as affected by earthworms, tillage and residues. *Soil Science Society of America Journal* 51, 1580–1586.

Discussion

Nowland: There is much more to biodiversity than the broad effects of the biota on soil. The diverse biota, whether increasing or decreasing, can be used to 'glue-down' soil by means such as by-products and exudates from microorganisms. This could reduce the urgency for the more mechanical methods of soil conservation, such as terraces and tillage practices. What are the prospects for using biodiversity to conserve soil?

Lal: I think that could be a good approach, especially if built into agronomic systems, such as the use of legume crops, mixed-cropping systems, and organic wastes, all of which would stimulate the microbial biodiversity of the soil. Non-mechanical systems are certainly to be preferred.

Hadley: A great deal is known about earthworms compared with other components of the soil fauna. Three functional groups were identified, including quick-impact beneficial species, and a pantropical species able to lock up nutrients that otherwise might be lost. Given these characteristics, how close are we to producing technological packages based on earthworms for use in tropical agriculture? Further, does not the nearness of that package based on low diversity show that low diversity can be beneficial to agriculture?

Lee: The problem is that we don't know enough about the effects of other species. On the present evidence we can't say whether with more species an even better result might be obtained. There is great interest in a technological package at this time, but more need for experimental work and calculations on the lines of those initiated by Darwin.

Nine

Soil Salinity and Biodiversity

I. Szabolcs, *Research Institute for Soil Science and Agricultural Chemistry of the Hungarian Academy of Sciences, P.O. Box 35, H-1525, Budapest, Hungary*

ABSTRACT: Salt-affected soils constitute a well-defined group of soil types including different soil formations with one common general characteristic, that they have higher electrolyte contents, both in their solid and liquid phases, than any other soil types. The high electrolyte content influences, directly or indirectly, all the substantial soil properties limiting soil fertility. Depending on the properties of the different salt-affected soils their biota is also diverse. Five practical groups of salt-affected soils are introduced and their biodiversity is described: saline, alkali, magnesium, gypsiferous and acid sulphate.

This contribution deals with the properties and biodiversity of the five groups of salt-affected soils, and also the possibilities for their utilization, as well as with considerations for the sustainability of agriculture and environment.

Introduction

Soil is an essential part of the ecosystem, and is in some respects a specific ecosystem itself, because it consists of both living and non-living materials, contains minerals, organic substances, animals, and micro- and macro-organisms. All these components play an important role in the mass and energy flow of soil-forming processes, and in the dynamics of plant nutrients in soils. The biota determine the productivity of soils and influence their sustainability as farmlands.

The different soil types represent, in a broad sense, specific patterns of the results of soil-forming processes, and each is associated with a certain type

The Biodiversity of Microorganisms and Invertebrates: Its Role in Sustainable Agriculture.
Edited by D.L. Hawksworth. © CAB International 1991.

I. Szabolcs

Table 9.1. World extent of salt-affected soils, according to continents.

Continent	Area (000s ha)
North America	15 755
Mexico and Central America	1 965
South America	129 163
Africa	80 538
South Asia	87 608
North and Central Asia	211 686
South-East Asia	19 983
Australasia	357 330
Europe	50 804
Total	954 832

of biota, characteristic in many respects of the given soil type. Consequently, the better we can describe the properties of a soil the better we can characterize its relation to the whole ecosystem in which the soil develops and exercises influence on the environment.

General characterization of salt-affected soils

Salt-affected soils are widely distributed and occur on all the continents, covering about 10% of their total territory. In Table 9.1, the areas of salt-affected soils on the continents (except Antarctica) are summarized (Szabolcs, 1989).

In spite of the fact that the term 'salt-affected soils' is unambiguous, it covers different soil formations. The soil types belonging to this large group have diverse physical, chemical, and biological properties and consequently different degrees of fertility and different biotas.

The only common feature of all salt-affected soils is that their soil solution has an electrolyte content, higher than any of the other soil types, and this high electrolyte content influences, either directly or indirectly, all the substantial soil properties limiting their fertility. At the same time, the concentration of electrolytes in the soil solution varies in a wide range, and the chemical nature of the electrolytes can also be very different in the different groups of salt-affected soils. Consequently, the effect of electrolytes is different on soil fertility, as well as on the types of invertebrates, micro-organisms, and flora living in or on the given soil.

The more or less high electrolyte content results in different physical, chemical and biochemical processes, and forms different morphological features, as well as different patterns of biota and nutrient dynamics in the soils.

Classification of salt-affected soils and their biota

In the various soil classification systems, salt-affected soils are placed in different classes and groups within the soil classification hierarchy. The principles and philosophy underlying the different systems are also diverse. This is the reason why the names in one soil classification system do not always have an equivalent in another. Another consequence of the diversity of soil classification systems is that their soil units are also associated with environmental conditions, with the biota, and with aspects of sustainable agriculture in different ways. This results in difficulties not only for practical users, but also for those who are not fully versed in the jargon of pedology. As a result, in many branches of science and technology, specific soil groupings or classification systems have been elaborated which are different from the orthodox pedological ones. Such groups are familiar to those concerned with the application of fertilizers, soil reclamation, irrigation, drainage, etc. The principles and systems used are suited to the target and aims of the practical user requirements for soil characterization.

To avoid the discussion and criticism of different soil classification systems, five practical groups of salt-affected soils are described here, paying due attention to their biodiversity. However, before presenting this practical grouping, which is applicable for different purposes, a few remarks should be made on the interrelation between salt-affected soils and their biota.

There is a misconception, not only among lay people but also in the technical literature, that salt-affected soils are 'abiotic' or they have very poor biota, but this is not so. Most salt-affected soils have a characteristic and well-developed biota, though in most cases they are not favourable for agri-, sylvi- or horticultural production. Consequently, the question should be approached from two directions: (i) the general characteristics of the biota in the different salt-affected soils; and (ii) the effect of the biota on soil productivity. The characteristics of the different soil types considered are provided in Table 9.2, and elaborated on below.

Saline soils

This group of soils is characterized by a high salt concentration and nearly neutral pH, which determine, together with other environmental factors, the specific biota of these soils.

Saline soils are the dominating group of salt-affected soils, found mainly in deserts and semi-deserts on all continents. The salt concentration in these soils is rather high and sometimes exceeds a few per cent of the total weight of the soil. Consequently, they occur not only in the soil solution but in many cases also on the solid face of the soils as crystalline salts. The salts consist mainly of sodium chloride and sodium sulphate; or sometimes even sodium

Table 9.2. Grouping of salt-affected soils and their properties limiting the biota.

Type of salt-affected soils	Electrolyte(s) causing salinity and/or alkalinity	Environment	Properties adversely affecting the biota	Method for reclamation
Saline soils	Sodium chloride and sulphate (in extreme cases – nitrate)	Arid and semi-arid	High osmotic pressure of soil solution, toxic effect of chlorides	Removal of excess salt (leaching)
Alkali soils	Sodium ions capable of alkaline hydrolysis	Semi-arid Semi-humid Humid	High alkali pH Adverse effect of poor water, physical soil properties, calcium deficiency	Lowering or neutralizing the high pH by chemical amendments
Magnesium soils	Magnesium ions	Semi-arid Semi-humid	Toxic effect, high osmotic pressure, calcium deficiency	Chemical amendments Leaching
Gypsiferous soils	Calcium ions (mainly calcium sulphate)	Semi-arid Arid	Acid pH, toxic effect, nutrient deficiency	Alkaline amendments
Acid sulphate soils	Ferric and aluminium ions (mainly sulphates)	Seashores, lagoons with heavy, sulphate-containing sediments Deluvial inland slopes and depressions	High acidity, toxic effect of aluminium, nutrient deficiency	Liming

nitrate. As a result, the pH of these soils is nearly neutral. The high salt concentration coagulates the colloids which results in comparatively favourable water-physical properties, making leaching possible.

Saline soils can be found, in addition to arid areas, on sea-shores and by lagoons as a consequence of salt accumulation from seawater. The salts are often transported by the wind.

These two types of situations where saline soils occur result in differences in their biota. In inland deserts, only a very poor flora and fauna can develop on these soils, comprising mainly halophytes and invertebrates. Some microorganisms can also be found, but not in great diversity and only in the upper soil layers. Owing to the austere conditions limiting the biota the less demanding species may survive and develop, for example algae, seaweeds, mastigomycetes and actinomycete bacteria.

In maritime saline soils, the biota is influenced by that of the seawater, sea-shore, or lagoon. Consequently, it differs in many respects from the biota of inland saline soils.

Considerable areas of saline soils are under cultivation as a result of amelioration, mainly drainage and irrigation. Not only field-crops but also vegetable and orchard plantations can be found in such areas, which contribute significantly to global food production, particularly in arid areas. The biota of saline soils alters completely with the introduction of intensive agricultural or horticultural production. Not only with respect to the cultivated crops, but also to the fauna and flora, and particularly the microorganisms, which change on ameliorated saline soils. Cultivation is associated with the appearance of more exacting species which dominate the biota in the oases. It should be noted, however, that the well-known hazards of salinity and waterlogging of irrigated soils also threaten the soil biota. If those adverse processes develop, not only is the sustainability of agriculture impaired, but the soil biota also changes in a direction unfavourable to the retention of useful soil properties and for agricultural production.

The activity of different microorganisms in saline soils has not been widely studied, but available sources show that they are of significant importance in some processes of soil formation and in the mass and energy flow of the soils. Karlson and Frankenberger (1990) reported that in salinized soils the microbial volatilization of selenium proved to be a potential bioremediation method for saline soils containing sodium sulphate and sodium chloride.

The incubation of ameliorated saline soils with *Rhizobium* strains is a common method of improving the biological activity of such soils when modern agrotechnics are introduced. Increased nitrogen fixation by this bacterium contributes to the development of microbiological communities, which are favourable to soil productivity. Improving the nutrient status and microflora of soils contributes substantially to the development of invertebrates and other soil-living animals which take part in the metabolism of soil substances, and so make them available for biological processes.

Alkali soils

Of all the groups of salt-affected soils, the alkali soils are markedly the most diverse in respect of their occurrence in different environmental conditions, as well as in their chemical and morphological properties and the diversity of their biota. Alkali soils are more 'cosmopolitan' than saline soils, and can be found practically in all climatic belts from the equator to inside the Arctic Circle. In Europe, alkali soils dominate among the salt-affected soils, in contrast to all other continents where, in the family of salt-affected soils, it is the saline soils which prevail.

Alkali soils develop mainly on heavy parent materials which can be subdivided, according to their clay mineral contents, into two groups: (i) the illite vermiculite type; and (ii) the montmorillonite type. As a rule, in alkaline soils the pH is high and the salt content is much lower than in saline soils.

The alkaline pH is mainly responsible in this type of salt-affected soil for the poor physical and water-physical soil properties. As a result of a high pH and comparatively low salt concentration, the soil colloids are dispergated, the porosity and structure of the soil are very poor, and the water availability for plants drops – drastically reducing nutrient uptake. Free sodium carbonate, which is toxic to plants, often occurs in alkali soils, and together with their often high boron contents (in addition to the poor physical composition of the soil), chemically prevents their agricultural utilization. It is well-known that the so-called exchangeable sodium content is high in alkaline soils; indeed, this is one of the diagnostic features of this group. Besides exchangeable sodium, the values of exchangeable magnesium are also often high (see p. 112).

Another difference between alkali and saline soils is that, in consequence of the lower salt concentration in the former, crop production is often possible without irrigation and even without chemical reclamation. In such cases the proper selection of fertilizers is essential because, evidently, in alkaline soils the fertilizers with an acidic reaction are much more effective than those of an alkaline one.

In relation to the soil fauna, flora, and microorganisms, alkali soils can be subdivided into two groups: (i) alkali soils with a high sodium carbonate content and high pH in which a columnar or crusty horizon B is not apparent; and (ii) alkali soils with a very low or no sodium carbonate content in the upper layers, where a diagnostic high horizon B develops with a columnar or crusty structure. All the solonetz and solod soils belong to this type.

These two types of alkali soils are inhabited by different organisms. As the tolerance of living things is very limited by the high sodium carbonate contents and the very high pH, which often exceeds 10 or 11, the flora of such soils consists of *Salsola soda* and related halophytes. Only invertebrates and microorganisms which can endure the toxic effects of soda survive. This situation is rather different in solonetz soils where the biota depends largely on the variety of the soil in question. Notably, if the horizon B with a high exchangeable sodium percentage is on, or near, the surface of the soil, the biota is very poor. However, where the horizon B is deep enough (20–30 cm or more) and is overlain by a horizon A (whose properties are always better than that of the horizon B), the biota of this upper layer will not differ too much from that of the surrounding non-salt-affected soils. They have a fairly rich flora, fauna, and microorganism content, with different invertebrate and microorganism communities capable of furthering processes useful for the nutrient metabolism for both plant roots and microbes.

The biological processes in alkaline soils have been well-studied and interpreted not only in respect of the biological communities of the soil, but also in respect of soil formation. Numerous publications (Szabó *et al.*, 1958) describe the ecology of *Streptomyces griseus* in Hungarian solonetz soils, and determined the spectrum of carbon and nitrogen utilization by their communities in relation to salt tolerance, optimum temperatures and pH, antibiotic activity, etc. In degraded solonetz soils, actinomycetes are predominant among the microorganisms and amount up to 40% of the whole microflora in the A horizon, and up to 75% in the B horizon.

Streptomyces griseus is one of the most significant organisms in this soil type as far as salt tolerance is concerned. *Streptomyces sterilis* strains, however, can tolerate even higher levels of sodium carbonate than those of *S. griseus*. The prevalence of *Streptomyces* strains in many solonetz soils evidently exercises great influence not only on the microbiological communities of the soils, but also on the invertebrates and other communities.

In solonetz and closely related solod soils, the organisms play a significant role in the synthesis and decomposition of soil organic material. Due to the alkaline conditions and changing moisture content in such soils, in the chain of synthesis and decomposition of organic mineral materials, different organisms have a role. In solonetz and solod soils, on the surface colonies of cyanobacteria, mainly *Nostoc commune*, form dark spots characteristic of solod-forming processes.

Another significant phenomenon is the influence of microorganisms on the formation of sodium carbonate in solonetz soils as a result of the soil-forming processes. As seen from the technical literature, there are numerous ways in which sodium carbonate is formed in soils, some consisting mainly of biological, and others of abiotic, processes. One of the patterns of sodium carbonate formation in soils is associated with the activity of biological sulphate reduction. This process is widespread in swamps or temporarily waterlogged areas, and requires a certain quality and quantity of organic matter in the soil for the activity of *Desulfovibrio desulfuricans* (Timar and Szabolcs, 1964). The formation of sodium carbonate can be described schematically by the following equations:

$$X_n SO_4 + 4H_2 = X_n S + 4H_2O$$

$$2CH_3 \cdot CH \cdot OH \cdot COONa + Na_2SO_4 = 2CH_3 \cdot COONa + Na_2S + 2H_2CO_3$$

$$Na_2S + H_2CO_3 \rightleftharpoons H_2S + Na_2CO_3$$

It should be noted that a particular amount of sulphate ions is a precondition of the above process. If the salt concentration is higher or lower, *Desulfovibrio desulfuricans* cannot function and the biological formation of sodium carbonate will not take place. In places where this process develops, the necessary concentration of sulphate ions are available, as a rule, for biological soda formation.

The two examples described clearly show that the biota and biological processes in alkaline soils play a very important role in forming the different soil varieties, and also in the development of the soil properties.

Magnesium soils

Among all the types of salt-affected soils, magnesium soils are the most discussed, but our knowledge and experience of this type are insufficient. What is known is that several formations of magnesium soils occur, and that in some of them soluble magnesium salts, in others absorbed magnesium ions, and often clay minerals, rich in magnesium, result in the particular properties of this type of salt-affected soil (Darab, 1980). Magnesium soils are distributed mainly in semi-arid and semi-humid areas, and they often occur in association with alkali soils.

As a consequence of the fact that magnesium soils include diverse formations, their biota is also diverse. Where magnesium soils are associated with alkali soils, their biota does not differ too much from that of the alkali soils. Where the magnesium soils are associated with zonal soil types such as chernozem soils or meadow soils (smolnitzas), however, their biota is similar to that of the local soil. Two facts should, however, be noted: (i) the organisms which demand a high magnesium supply are always prevalent in magnesium soils; and (ii) if the soil contains high amounts of magnesium chloride, its toxic effect manifests itself in the spectrum of the organisms in this soil.

Gypsiferous soils

This group of soils represents yet another insufficiently studied type of salt-affected soil, although it covers large areas in arid- and semi-arid regions. It occurs in practically all continents, but is particularly frequent in North Africa, some western states of the USA, in Central Asia, in the Middle East, and the Far East.

The salt component of gypsiferous soils is dominated by calcium sulphate, which sometimes constitutes more than half of the total soil material.

Agricultural utilization of such soils is practically impossible, but a large proportion of the gypsiferous soils with a lower calcium sulphate content are suitable, and actually utilized for, irrigated agriculture. The main constraints of nutrient uptake from such soils, are: (i) acidic pH; (ii) toxic effects; and (iii) calcium versus potassium antagonism. The total and mobile calcium content of gypsiferous soils differs according to the parent material of the soils, particle size distribution, environmental conditions, and other factors.

The high gypsum content of these soils exercises the dominant influence on the soil biota. In natural vegetation the plants able to survive and develop require or can tolerate high gypsum contents. The same is true for the soil fauna and microorganisms.

In many parts of the world, gypsiferous soils are used for irrigated farming. In this respect, the same phenomena can be observed as are described under saline soils above (p. 109).

Acid sulphate soils

Acid sulphate soils constitute a specific group of salt-affected soils to be found in tidal marshes, lagoons, and sulphuric sea-shore sediments along the coastlines of all continents from Finland South to the Gulf of Guinea, and from Madagascar to Vietnam. In some countries, for example, Thailand and South India, vast territories are covered by this type of salt-affected soil (Szabolcs, 1989).

Inland acid sulphate soils can also be found in different areas of the world, such as the western territories of the USA, Asia Minor, and China. Such soils developed as a result of fluvial glacial processes and have no connection with seashores in recent geological times.

As a consequence of the nature of acid sulphate soils, and the high aluminium and iron sulphate content in their upper layer when they emerge from the sea, the pH of these soils can be as low as 2–3; sometimes even free sulphuric acid occurs. Evidently in this state they are unfit for any kind of growth, but after liming they are very good for the production of many crops, particularly, rice. The two dominating clay minerals in acid sulphate soils are pirite and jarosite (van Breemen, 1973).

In the formation of acid sulphate soils, the dominating process is the oxidation of bivalent sulphur into six-valent sulphur. This process takes place under the influence of oxygen in the air even in the absence of biological factors. It is interesting to note that the change in the valence of the sulphur has a significant role in the opposite way in another group of salt-affected soils, namely alkali soils during the biological formation of sodium carbonate (pp. 111–112).

Under natural conditions, *Desulfovibrio desulfotomaculum* often also occurs during the formation of these soils, because in the sulphur cycle diverse processes may develop. Some of the sporulating strains of this species can exist in their vegetative stage under extreme pH conditions (Rickar, 1973). Microorganisms also take part in the oxidation of sulphur. Most of such microorganisms belong to *Thiobacillus*, a genus of chemoautotrophic organisms capable of utilizing the energy obtained from the oxidation of hydrogen sulphide, sulphur, sulphite, bisulphite and trisulphide to sulphate for the assimilation of carbon dioxide. In general they are aerobic, but *Thiobacillus denitrificans*, and perhaps *T. thioparus*, can grow anaerobically if nitrate is present as an electron acceptor. *Thiobacillus thiooxidans* and *T. ferrooxidans* are remarkable for their tolerance of extreme metal iron concentrations and very acid conditions. *T. ferrooxidans* and *Ferrobacillus ferrooxidans* are able to mediate the oxidation of ferrous to ferric ions. Members of the Beggiatoaceae, photoautotrophic bacteria of the Rhodobacteriae, and *Sphaerotibus* can also catalyze the oxidation of sulphur compounds, but *Thiobacillus* is by far the most important (van Bremen, 1973).

As far as acid sulphate soils occur in areas with different climatic and geographical situations, their biota can be diverse in close correspondence to the environmental conditions. In seashores, lagoons and tidal marshes, the maritime type of flora and fauna prevails, while in inland acid sulphate soils this is lacking and another type of biota occurs.

It is well-known that acid sulphate soils are well-developed in mangrove ecosystems, whose biota is determined by such circumstances. Reed vegetation in brackish swamps is also associated with acid sulphate soils, for example in West Africa where racemose is common, or in South America where *Rhizophora mangle* has been studied (Brinkman and Pons, 1973). In freshwater areas, *Imperate brasiliensis* (white cedar) is characteristic of South America, in eastern Asia *Melaleuca* can be observed. The latter tree has a wide ecological amplitude.

The invertebrates and other members of the fauna are strongly influenced by local climatic and hydrological conditions.

Many of the acid sulphate soils of the world are under cultivation, mainly for rice production. It is evident that rice-growing is possible only when the acidic pH is neutralized, or permanent waterlogging does not permit oxidation processes in tidal marshes, and acidity cannot increase. In cultivated acid sulphate soils the biota changes completely and becomes similar to that of other paddy fields.

Conclusions

This survey of the properties and biota of salt-affected soils shows that diverse types of biota occur depending both on the character of the different soils and on local environmental conditions.

In order to sustain the fertility of a soil and increase its productivity, a site-specific study is always necessary to characterize the biological communities and exploit the possibilities of its productive capacity. This may contribute to the better utilization of such soils. It must always be remembered that agriculture, horticulture, or sylviculture always alter the biota in the soils concerned. The alterations may be for better or for worse. If we can prevent harmful processes and further those favourable for production, knowledge of the dynamics of soil biology can contribute to sustainability and increase soil fertility.

References

van Breemen, N. (1973) Soil forming processes in acid sulphate soils. *Publications of the International Institute for Land Reclamation and Improvement, Wageningen* 18(1), 66–130.

Brinkman, H. and Pons, L.J. (1973) Recognition and prediction of acid sulphate soil conditions. *Publications of the International Institute for Land Reclamation and Improvement, Wageningen* 18(1), 169–203.

Darab, K. (1980) Magnesium in solonetz soils. In: *International Symposium on Salt Affected Soils*. Karnal, pp. 92–101.

Karlson, U. and Frankenberger, W.T. (1990) Alkylselenide production in salinized soils. *Soil Science* 149, 56–63.

Rickar, D.T. (1973) Acid sulphate soils. *Publications of the International Institute for Land Reclamation and Improvement, Wageningen* 18(1), 28–65.

Szabó, I., Marton, M. and Szabolcs, I. (1958) Adatok a *Streptomyces griseus* Waksman *et al.* ökológiájának ismeretéhez. *Agrokémia és Talajtan* 7, 163–176.

Szabolcs, I. (1989) *Salt Affected Soils*. CRC Press, Boca Raton, Florida.

Timár, M. and Szabolcs, I. (1964) Szervesanyag hatása a szikesekben folyó szulfátredukcióra. *Agrokémia és Talajtan* 13, 129–136.

Ten

Biodiversity and Sustainability of Wetland Rice Production: Role and Potential of Microorganisms and Invertebrates

P.A. Roger[1], K.L. Heong & P.S. Teng, *International Rice Research Institute, PO Box 933, Manila, Philippines*

ABSTRACT: Some of the ecological foundations of sustainable wetland rice production related to microorganisms and invertebrates and their biodiversity are considered including: (i) aspects of sustainability of rice-producing environments involving microbial and invertebrate populations, the maintenance of soil fertility, effects and control of rice pests and vector-borne diseases; (ii) how crop intensification affects these populations and their biodiversity; (iii) agricultural practices that use microbial and invertebrate populations and their biodiversity; and (iv) the status of germplasm collections and the potential of biotechnology to use them to improve the sustainability of rice-producing environments.

The beneficial and detrimental roles of microorganisms and invertebrates in sustainable rice production have been identified and, sometimes, quantified. However, less is known about the possible long-term effects of crop intensification on these populations and their biodiversity. Numerous methods using microorganisms and invertebrates to increase soil fertility and control pests and diseases have been tested. But the success of these methods is limited and their adoption almost negligible. This will probably continue while the methods are still based on a very restricted knowledge of biodiversity, community structure, and trophic relationships at the ecosystem level.

Recent data on arthropods confirm that high biodiversity does not imply stability and low pest populations. Increases in pest and vector densities depend more on predator diversity, species resilience to perturbations, and biological attributes. Thus, increasing or preserving diversity *per se* does not necessarily contribute to pest stability, but developing effective trophic linkages might. This approach might also be valid for maintaining soil fertility through microbial management, the

[1] Present address: Laboratoire de Microbiologie ORSTOM, Université de Provence, Case 87, 3 Place Victor Hugo, F-13331 Marseille Cedex, France.

The Biodiversity of Microorganisms and Invertebrates: Its Role in Sustainable Agriculture.
Edited by D.L. Hawksworth. © CAB International 1991.

optimization of primary production in floodwater, and the optimization of nutrient recycling by invertebrate populations.

Introduction

More than half of the world's population depends on rice which, in 1988, occupied 145 million hectares of land, with a global production of 468 million tonnes. An additional 300 million tonnes of rice will be needed in 2020 to meet the need of a fast-growing human population. This requires a 65% production increase within 30 years without much expansion of the actual cultivated area (International Rice Research Institute, 1989). However, increased rice production should not be at the expense of future generations and should fulfil the concept of sustainability. It should be achieved through rice production management that: (i) satisfies changing human needs and maintains production over time in the face of ecological difficulties and social and economic pressure; (ii) maintains or enhances the quality of the environment; and (iii) conserves or enhances natural resources. Aside from maintaining growth in productive agricultural systems and promoting growth in less productive systems, the major issues are: (i) managing pests and nutrients in ways that reduce agrochemical use; (ii) preserving the natural resource base; and (iii) protecting the genetic base for agriculture.

This review considers the aspects of sustainability of rice production involving microbial and invertebrate populations, the effects of crop intensification on these populations, the agricultural practices that utilize microorganisms and invertebrates, and the current status of germplasm collections and their biotechnological use for improving the sustainability of rice-producing environments.

Importance of microorganisms and invertebrates in the sustainability of rice-producing environments

Maintenance of soil fertility by microorganisms and invertebrates

From the point of view of yield sustainability, traditional wetland rice cultivation has been extremely successful. A moderate but stable yield has been maintained for thousands of years without deterioration of the environment (Bray, 1986). This is because flooding favours soil fertility and rice production by: (i) bringing soil pH near to neutral; (ii) increasing the availability of nutrients, especially phosphorous and iron; (iii) depressing soil organic matter decomposition and thus, maintaining soil nitrogen fertility;

(iv) favouring nitrogen fixation; (v) depressing outbreaks of soil-borne diseases; (vi) supplying nutrients from irrigation water; (vii) depressing weed growth, especially those of the C-4 type; and (viii) preventing water percolation and soil erosion (Watanabe *et al.*, 1988).

There is, however, no assurance that, in the long-term, crop intensification will not affect wetland soil fertility. Research on rice nutrition has shown that, at usual levels of inorganic fertilizer applied to ricefields, most nitrogen absorbed by the plant originates from the soil, where it is released by the turnover of a microbial biomass representing only a small percentage of total soil nitrogen (Watanabe *et al.*, 1988). Crop residues, rhizosphere exudates, and photosynthetic aquatic biomass (algae and aquatic plants) contribute nutrients that allow microbial biomass replenishment. Crop residues are incorporated at the beginning of the cropping season while nutrients accumulating in the photosynthetic aquatic biomass (including biologically fixed atmospheric nitrogen) are continuously recycled and reincorporated into the soil by zooplankton and the soil fauna, which are therefore key components of ricefield fertility (Roger and Kurihara, 1988). Some of the inputs allowing the replenishment of microbial biomass have been quantified, but a comprehensive understanding of the mechanisms involved in this aspect of nitrogen cycling is still to be developed. It is important to understand and predict how factors associated with crop intensification (e.g. agrochemicals) may affect the soil microbial biomass, directly or indirectly, by decreasing the productivity of the photosynthetic aquatic biomass and the populations of invertebrates responsible for recycling soil nutrients.

Yield losses caused by microbial and invertebrate pests of rice

Of the approximately 100 insect species and 74 diseases and physiological disorders associated with rice (Teng, 1990), 30 insects and 16 diseases are considered economically important (Riessig *et al.*, 1986). Table 10.1 presents a summary of estimates of yield loss due to pests and diseases.

The important insect pests of wetland rice are brown planthopper, leaffolder, stem borer, and green leafhopper (a vector of tungro virus). Important but localized losses have been attributed to rice bug (*Leptocorisa* species), gall midge larvae (*Pachydiplosis oryzae*), rice hispa (*Dicladispa armigera*), and armyworm (*Mythimna separata*) (Teng, 1986, 1990). Scarce information exists on losses caused by nematodes. Important diseases are tungro virus, sheath blight, bacterial blight, and blast (on susceptible varieties). In recent years, tungro has become a major problem in many tropical areas because of its potential to cause total loss and the lack of corrective measures once its symptoms are observed. Few data are available on yield losses caused by other diseases; but generally, under favourable conditions, most pathogens have the potential to cause severe losses (Teng *et al.*, 1990).

Table 10.1. Estimates of losses due to rice pests (adapted from Teng, 1990).

Agent	% Loss	Location	Reference[*]
Insect pests			
All insects	24%	Asia	(Ahrens et al., 1982)
	35–44%		(Pathak and Dhaliwal, 1981)
	35%	India	(Way, 1976)
	16–30%	Philippines	(Way, 1976)
	6%	Bangladesh	(Alam, 1961)
	10–20%	Sri Lanka	(Fernando, 1966)
Rice stem borers	30–70%	Bangladesh[†]	(Alam et al., 1972)
	3–20%	Bangladesh[‡]	(Alam, 1961)
	3–95%	India	(Ghose et al., 1960)
	Up to 95%	Indonesia	(Soenardi, 1967)
	33%	Malaysia	(Wyatt, 1957)
Leafhoppers	50–80%	Bangladesh	(Alam, 1961)
Brown planthopper	1–33%	India	(Jeyaraj et al., 1974)
Rice bugs Leptocorisa	10%		(Pruthi, 1953)
Gall midge larvae	12–35%	India	(Reddy, 1967)
Pachydiplosis oryzae	50–100%	Vietnam	(Reddy, 1967)
Rice hispa	10–65%	Bangladesh	(Barr et al., 1975)
Dicladispa armigera	Up to 50%	India	(Barr et al., 1975)
Leaf-folders	Up to 50%	India	(Balasubramaniam et al., 1973)
Diseases			
Blast	Up to 100%		(Teng et al., 1990)
	5–10%	India	(Padmanabhan, 1965)
	3%	Japan	(Teng, 1990)
	8–14%	China	(Teng, 1986)
Brown spot	80%	India	(Padmanabhan, 1973)
	14–41%	India	(Vidhyasekaran and Ramados, 1973)
Sheath blight	9–13%	China	(Teng, 1986)
Tungro virus	100%	Indonesia	(Chang et al., 1985)
	40–60%	Bangladesh	(Reddy, 1973)
	50%	Thailand	(Wathanakul and Weerapat, 1969)
	30%	Philippines	(Teng, 1990)
Bacterial blight	Up to 60%	India	(Srivastava, 1972)
	5–6%	China	(Teng, 1986)
Stem rot	5–10%	India	(Chauhan et al., 1968)
	5–6%	China	(Teng, 1986)

[*]Bibliographic details of references are listed in Teng (1990); [†]outbreak; [‡]chronic.

The generalized crop loss figures most commonly cited are those by Cramer (1967), who concluded that more of the rice potential production is lost due to pests (55%) than is harvested (45%). He estimated that the percentage of the potential harvest lost due to pests was 34% due to insects, 10% to diseases, and 11% to weeds. Although these figures appear to be high,

Teng (1990) found that other authors have felt that there is no sound evidence to the contrary and that these values may be underestimations of actual losses in some years. Other generalized estimates for losses caused by insects of tropical rice are 35–44% (Pathak and Dhaliwal, 1981), 24% in East and South-East Asia (Ahrens *et al.*, 1982), 35% in India, and 16–30% in the Philippines (Way, 1976).

Vector-borne diseases

Wetland rice culture and irrigation schemes in tropical and subtropical regions create ecological conditions favourable to the propagation of vector-borne diseases. The most important of these are malaria, schistosomiasis, and Japanese encephalitis, whose vectors require an aquatic environment. The invertebrate vectors of human diseases in rice-growing environments are basically mosquitoes and aquatic snails.

The reproduction of mosquitoes in ricefields is affected by plant height, water depth, soil and other environmental conditions, and cultural practices. Generally, larval populations are low after transplanting, peak a few weeks later, and decline as the plants reach a height of 60–100 cm. Mosquito reproduction in ricefields ranges from 2 to $20\,\mathrm{m}^{-2}\,\mathrm{day}^{-1}$ (Roger and Bhuiyan, 1990).

Aquatic snails are very common in ricefields where they can develop large populations, especially at the beginning of the cropping season when organic manure is applied. Populations up to $1000\,\mathrm{m}^{-2}$ have been observed in Philippine ricefields. Behavioural experiments showed that snails having to choose between various soils were most often attracted (75%) to rice.

Effects of crop intensification on microbial and invertebrate populations

General effects of crop intensification on biodiversity in ricefields

Traditional ricefields, some of which have been cultivated for several hundred years, may be considered as climax communities. Modern technologies, which utilize fertilizer-responsive varieties, fertilizers, pesticides, and optimum water and crop management practices, have tremendously increased yields and production but have, indeed, caused profound modifications to traditional rice-growing environments.

In general, a disturbance to a stabilized ecosystem reduces the number of species while provoking 'blooms' of certain others; such effects have been observed in ricefields (Roger and Kurihara, 1988). However, quantitative knowledge of the long-term effects of crop intensification on species diversity

Table 10.2. Summary of quantitative records of species/taxa in wetland ricefields.

1. Number of species recorded by Heckman in 1975 in a 1-year study of a single field in north-eastern Thailand (six samplings)

Sarcodina	31	Cyanobacteria	11
Ciliata	83	Algae	166
Rotifers	50	Pteridophyta	3
Platyhelminths	7	Monocotyledonae	25
Nematoda	7	Dicotyledonae	10
Annelida	11	Pisces	18
Mollusca	12	Amphibia/Reptilia	10
Arthropoda	146		
		Total	590

2. Number of species/taxa of aquatic invertebrates, excluding protozoa, recorded by different authors
- Heckman (1979) (species), one traditional field, 1-year study (Thailand)   183
- Lim (1980) (taxa), 2-year study of pesticide application (Malaysia)   39
- Takahashi *et al.* (1982) (taxa) four fields, single samplings (California)   10–21
- International Rice Research Institute (1985) and Roger *et al.* (1985) (species) single samplings in 18 fields with pesticide applied (Philippines and India)   2–26

3. Records of arthropod species in ricefields over one crop cycle
- Kobayashi *et al.* (1973): study in 1954–55 of several fields by net sweeping (Shikoku, Japan)   450
- Heong *et al.* (unpubl. data): study in 1989 of five ricefields by suction (Philippines):   Fields considered separately: 146, 125, 116, 92, 87
                                                   Five fields combined   240

is extremely scarce. The only reference on the species abundance in traditional ricefields is a 1975 study by Heckman (1979) in Thailand, where 590 species (excluding fungi) were recorded in one field within 1 year (Table 10.2). Few records of aquatic invertebrates can be compared with Heckman's record of 183 species (Table 10.2). In a 2-year study of pesticide applications on Malaysian ricefields, Lim (1980) recorded 39 taxa of aquatic invertebrates. Single sampling by Takahashi *et al.* (1982) in four Californian ricefields recorded 10–21 taxa. In 18 sites in the Philippines and India, the highest number of aquatic invertebrate taxa recorded by single sampling at one site was 26, and the lowest 2 (Roger *et al.*, 1987). Similarly, records of numbers of arthropod species in Japanese ricefields estimated in 1954–1955 by net sweeping (Kobayashi *et al.*, 1973) seem to indicate a higher biodiversity than in recent data collected by Heong *et al.* (unpublished data) in five fields in the Philippines using the suction method (Table 10.2). All the above data were obtained using different sampling methods and time frames. The marked decrease of values recorded since 1975 might probably be taken as a rough indication of a decrease in total number of species after crop intensification; however, this does not demonstrate the generally accepted concept that crop intensification decreases biodiversity in ricefields.

Crop intensification has reduced the number of edible species traditionally harvested from ricefields. Heckman (1979) reported that one vegetable and 16 edible animal species (snail, prawn, crab, large water bug, fish and frog) were collected in a single ricefield within 1 year. Such a diversity is not common anymore, and pesticides may have rendered these edible species unfit for human consumption.

Agrochemical use, besides increasing rice yield, may also cause uncontrolled growth of single species that might, directly or indirectly, have detrimental effects. One of those effects is the outbreak of pests (Heinrichs, 1988) and other organisms that may affect the fertility- or health-related aspects of the ecosystem, for example: (i) blooms of unicellular algae, observed after fertilizer application, which cause nitrogen losses by volatilization; (ii) proliferation of ostracods and chironomid larvae, observed after insecticide application, which inhibits the development of efficient nitrogen-fixing blue-green cyanobacterial blooms; and (iii) proliferation of snails or mosquito larvae that may occur after insecticide application and favour vector-borne diseases (Roger and Kurihara, 1988).

Effects on soil and water microbial populations

Most of the information on the impacts of crop intensification on the ricefield microflora concerns pesticide use that may: (i) alter activities related to soil fertility; and (ii) reduce pesticide efficiency because of shifts in microbial populations toward organisms more efficient in their degradation. More than 200 papers, reviewed by Roger (1990), have been published on this topic, but more than half of the studies are short-term laboratory experiments in test-tubes or flasks that cannot be extrapolated to field conditions. Field and long-term laboratory studies on soil with pesticide levels near the recommended field dose allow us to draw the following conclusions.

Pesticides have three major effects on ricefield algae: (i) a selective toxicity that affects preferentially green algae and thus promotes cyanobacterial growth; (ii) a short-term promoting effect of insecticides on microalgae caused by a temporary decrease of invertebrates that graze on algae; (iii) a selective effect of insecticides on the cyanobacterial flora by causing a recruitment of grazers which results in the dominance of strains forming mucilaginous macrocolonies (e.g. *Nostoc*) resistant to grazing.

    Field and laboratory studies showed that pesticides applied to soil at the
recommended rates and intervals had either no effect on microbial populations or their activities, or had an effect that was followed by recovery after
1–3 weeks. Herbicides seem to have more short-term negative effects on the
soil microflora than insecticides. A few studies indicate that repeated applications of the same pesticide may cause its rapid inactivation because of the
enhanced growth of related specific decomposing microorganisms. This was
observed in gamma-BHC, diazinon, aldicarb, and nitrophenols, but not in
carbofuran and benthiocarb. Repeated application of a pesticide may also
change the metabolic pattern of its decomposition. In the case of benthiocarb
such a change produced a very phytotoxic compound (Moon and Kuwatsuka,
1984).
    Because of the lack of field studies over several crop cycles, there is no
information on the long-term effects of pesticide use on the wetland rice soil
microflora. No method is yet available to quantify the biodiversity of the soil
microflora.

Effects on invertebrate populations

Studies of the effects of pesticides on floodwater populations show that
insecticides are usually the most active compounds. Their application usually
causes a general decrease in floodwater invertebrates, followed by the
proliferation of primary consumers, notably ostracods, chironomid and
mosquito larvae, and molluscs (Ishibashi and Ito, 1981; Roger and Kurihara,
1988), while populations of predators such as odonate larvae are reduced
(Takamura and Yasuno, 1986). The rapid recovery of ostracods after
pesticide application results from their resistance to pesticides and the large
number of eggs they produce parthenogenetically (Lim and Wong, 1986).
    Nematodes and oligochaetes are probably the only soil invertebrates
studied in wetland ricefields. Usually, the specific abundance of parasitic
nematodes is higher in wetlands than in uplands, but apparently this results
from submersion rather than from higher agrochemical use in wetlands
(J.C. Prot, personal communication). Benthiocarb had no marked effect on
the number of nematode species and their average populations during the
crop cycle (Ishibashi and Ito, 1981). Studies at the International Rice
Research Institute showed a 70% reduction in soil oligochaete populations
when the amount of Furadan applied was increased from 0.1 to
1.5 kg a.i. ha^{-1} (I. Simpson, personal communication).

Effects on rice pests

The effects of new rice technologies on the carrying capacity of the ecosystem
for insect pests were summarized by Heinrichs (1988). The availability of

short-duration varieties and irrigation water has made rice cultivation throughout the year possible, thus eliminating fallow periods that often depress insect pests. Stable water supply has favoured aquatic pests such as the caseworm *Nymphula depunctalis* (Heinrichs and Viajante, 1987). Increased nitrogen fertilizer use on responsive new varieties has favoured the brown planthopper (Denno and Roderick, 1990). In general, BPH survive better, moult into larger adults, and are more fecund if they develop on nitrogen-rich host plants. The increased yield potential of modern varieties has also resulted in the misconception by farmers that greater returns will arise from pesticide application. Many agricultural authorities have thus subsidized and encouraged insecticide use. This in turn resulted in pesticide misuse, accelerated development of resistance in pests, destruction of natural control, and pest resurgence and outbreaks.

The effects of rice production intensification on microbial pests have been summarized by Teng (1990). Crop intensification has generally resulted in increased prevalence, incidence, and severity of diseases caused by bacteria, viruses, and fungi. Bacterial blight (*Xanthomonas campestris* pv. *oryzae*) and sheath blight (*Rhizoctonia solani*) are directly attributable to cultural conditions of the modern high-yielding rice varieties which are grown with nitrogen fertilizers in large homogeneous areas. However, improved fertility associated with crop intensification has also resulted in the decrease of diseases such as brown spot (*Bipolaris oryzae*). In areas with inefficient irrigation schemes, growing several crops a year has resulted in large areas with asynchronously planted rice, which is known to favour the devastating tungro virus. Disease epidemics cause instability in rice production over time because of the pathogen's ability to overcome resistance incorporated into the rice varieties.

There is new evidence that crop intensification has no significant effect on the diversity of pathogen species in tropical rice. A study of 90 fields in the Philippines (Elazegui *et al.*, 1990) showed that the number of pesticide and nitrogen fertilizer applications had no effect on the average number of pathogenic species encountered in the fields. However, transplanted rice was richer in pathogenic species and more diverse than directly seeded rice. A denser plant population might be less conducive for pathogen dispersal within a field.

Effects on vector-borne diseases

In traditional ricefields, although many vectors exist in the ecosystem, competition and predator pressure by fish and aquatic insects limit the productivity of any one vector. Ricefields contain a variety of insect predators of mosquito larvae such as backswimmers, gerrids, etc. (Hemiptera, Notonectidae), dragonfly and damselfly nymphs (Odonata), and adult and larval predacious water beetles (Coleoptera, Dytiscidae) (Service, 1977). Predator fauna vary according to rice cultivars, plant height, and water management (Mather and Trinh Ton That, 1984).

In rice monoculture, with a less diverse fauna and without control measures, the productivity of some vectors may be very high. Insecticides used to control rice pests and vectors may create secondary problems. The three major effects of insecticides (Roger and Bhuyian, 1990) are: (i) the temporary decrease in vector incidence since many agricultural insecticides are non-specific and affect some vectors, extensive agricultural insecticide use probably explaining the marked reduction of malaria and Japanese encephalitis in Japan after 1945; (ii) the resurgence of resistant strains, 50 malaria vectors resistant to one or more pesticides were recorded in the world in 1987; and (iii) the adverse effects on the natural predators and competitors of vectors, causing blooming of mosquito larvae, and molluscs which are usually not affected by most rice pesticides and which multiply because of reduced predation or competition for food.

Use of microbial and invertebrate biodiversity to enhance agricultural sustainability of wetland soils

Microbial management of wetland rice soils

The microbial management of wetland soils was reviewed by Roger *et al.* (1991). Using biological nitrogen fixation as an alternative or supplementary nitrogen source for rice has been the major approach. Whereas nitrogen-fixing green manures (*Azolla* and legumes) have been used for centuries in some rice-growing areas, research on nitrogen-fixing cyanobacteria and bacterial inoculants for wetland rice is relatively recent, being initiated in the early 1950s for cyanobacteria and in the 1960s for other bacteria.

Biomass estimates, nitrogen-fixation measurements, and inoculation experiments indicate that cyanobacteria, as an additional nitrogen source for rice, have a potential of 20–$30\,kg\,N\,ha^{-1}\,crop^{-1}$ which may translate to a yield increase of 200–$350\,kg\,ha^{-1}$. Recent data show that cyanobacteria are ubiquitous in rice soils and that foreign strains usually do not become established in the field. Thus, the principle of cyanobacterial inoculation should be reconsidered and more attention should be paid to promoting indigenous strains.

Reported effects of the bacterial inoculation of rice have been inconsistent. Most strains tested have been nitrogen-fixing forms, but there was no clear evidence that promotion of rice growth and nitrogen uptake was due to increased biological atmospheric nitrogen fixation. Therefore, several authors refer to the production of plant growth regulators to explain the beneficial effect of bacterial inoculation. No experiment has yet supported this hypothesis. The few data on strain establishment show that, in most cases, inoculated strains do not multiply. Given the current status of our knowledge on the bacterial inoculation of rice, no positive conclusion can be drawn as to its potential.

There are several reports on the existence of varietal differences in the ability to support associative biological nitrogen fixation (*Nfs* character). The idea of breeding varieties with higher nitrogen-fixing potential (*Nfs*) is attractive since it would enhance biological nitrogen fixation without additional cultural practices. This promising approach is still limited by the lack of an efficient screening method.

Use of biodiversity to control insect pests

Natural parasitoids, predators, and pathogens that attack insect pests are abundant in ricefields. Outbreaks occur when the equilibrium is disrupted and the full reproductive capacities of pest species are released. Thus, enhancing the action of natural control agents is of paramount importance in an integrated pest management (IPM) programme. The various aspects of IPM in Asia were reviewed by Teng and Heong (1988).

Pesticides can markedly affect the natural enemy fauna and should be used judiciously. In a study comparing 330 crops in insecticide-treated and untreated fields, only 50% of the fields showed measurable yield losses due to pests (Litsinger, 1984). The removal of pesticide subsidies in Indonesia led to a drastic decrease in pesticide use with no measurable reduction in the national yield average (Kenmore, 1989). Recent data from the Philippines also showed no significant difference in average yields between farms using and not using pesticides (Elazegui *et al.*, 1990).

IPM developed from the concept of integrating control tactics into an acceptable system. These tactics include ways to maximize natural control, and the use of resistant varieties and chemicals only when necessary. It is sometimes defined as the farmers' 'best mix' of control tactics based on crop yield, profit, and safety (Kenmore *et al.*, 1985). Since IPM advocates the conservation of natural enemy populations in ricefields, more basic knowledge on population dynamics and trophic linkages is needed. Increasing general biodiversity *per se* may not be sufficient to promote ecosystem stability and sustain low-pest situations. A recent study of arthropod diversity in Philippine ricefields has confirmed this hypothesis. Of the five locations where samples were collected, the International Rice Research Institute experimental farm had significantly higher biodiversity (Fig. 10.1), but the pest population was also significantly higher (Table 10.3). The lowest biodiversity and phytophage populations were recorded in traditional rice terraces.

Mean nº of species per sample

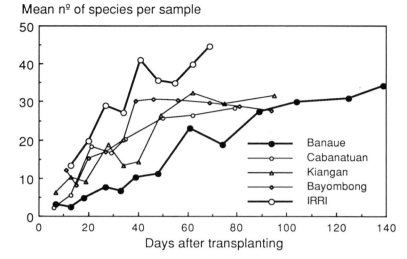

Figure 10.1. Dynamics of arthropod diversity during a crop cycle in five ricefields in the Philippines. (Banaue: rice terrace where rice has been grown without agrochemicals for centuries. Cabanatuan, Kiangan, Bayombong: farmers' fields with agrochemical use. International Rice Research Institute (IRRI): Experimental plot on the IRRI farm.)

Use of biodiversity to control rice diseases

Natural control is not as effective with pathogens as with insect pests, but there is evidence that antagonistic bacteria can control some of the agents of rice diseases. Rosales *et al.* (1968) found that 60% of 139 ricefield bacterial isolates inhibited the *in vitro* growth of *Fusarium moniliforme* (agent of bakanae disease). Disease control by seed treatment of 18 bacterial isolates tested in artificially infested nurseries ranged from 70 to 96%. Mew and Rosales (1986) reported reductions in sheath blight severity by bacterization with *Pseudomonas* and *Bacillus* strains. These methods are not currently used in the field.

Spatial and temporal diversity of rice is a well-known strategy for disease control. Blast has been controlled by using variety mixtures (Bonman *et al.*, 1986) and one control of tungro is by rotation of host genes conferring resistance (Manwan *et al.*, 1987). The principle for using host genetic diversity to control pathogens depends on an understanding of the population genetics of the pathogen and the frequency of virulence genes in different ecosystems. While much of the theory has been developed, few empirical data yet exist in the tropics to extend the concept to pathogens other than blast and tungro.

Table 10.3. Arthropod community structure in five ricefields in the Philippines.*

Sampling site	International Rice Research Institute	Cabanatuan	Bayombong	Kiangan	Banaue
Nature	Experimental farm	Farmer's fields			Rice terrace
Species number	31	16	20	17	8
Abundant species number	13	5	11	10	6
% Contribution	82	80	60	73	59
Evenness	0.58	0.52	0.62	0.70	0.86
Total number × 1000	22	15	8	11	6
% Phytophages	64	45	43	45	57
% Predators	26	53	52	51	35

*Average values per sampling; from Heong *et al.* (unpublished).

Use of biodiversity to control vector-borne diseases

The biological control of vectors basically has two major approaches: (i) maintaining species diversity and thus conserving natural predators; and (ii) introducing new predators, competitors, parasites, or diseases of vectors.

Most of the information on the conservation of natural predators of vectors refers to mosquitoes (Roger and Bhuiyan, 1990). Because of predators such as fish, Odonata, Notonectidae, and Discidae, the survival percentage of mosquito larvae in ricefields from the first-instar through the pupal stage varies from 2 to 5%. Spiders also reduce the number of adult mosquitoes. Despite the very high predation on larvae, there may still be large numbers of adults emerging and constituting a nuisance or disease hazard. If natural predators are destroyed, emerging mosquitoes are likely to be more numerous. Even if it is not envisaged that predators will be used as control agents, cultural practices favouring their existence should be encouraged (Mather and Trinh Ton That, 1984).

Numerous competitor and predator species have been tested to control vectors. The most promising method for mosquito control is to stock food fish in and around ricefields. It reduces vector and weed incidence, increases rice yields, partly because of the fish excreta, and produces fish food (Self, 1987). In the Philippines, the combined culture of larvivorous *Tilapia* and common carp in ricefields, with supplemental feeding, produced about 700 kg fish ha^{-1} year^{-1} (Petr, 1987). However, experience in the efficiency of introduced larvivorous exotic fish has been varied (Roger and Bhuiyan, 1990). The main constraints to ricefield fisheries and vector control by fish are the toxicity of agrochemicals, especially pesticides, and unreliable water supply.

The use of competitors is a strategy that seems to have been restricted to snail vectors. Large snails such as *Marisa* and *Thiara* were successfully introduced and supplanted schistosome vector snails in ponds and canals in several countries, but some of these large snails feed on transplanted rice seedlings (Roger and Bhuiyan, 1990).

Microbial agents with a potential to control mosquitoes include viruses, bacteria, and fungi. *Bacillus thuringiensis* serotype H-14 and several virulent strains of *Bacillus sphaericus* (Dame *et al.*, 1988) provide selective control of mosquito larvae, while causing relatively little harm to most of the predators of vectors and agricultural pests. Currently, only *B. thuringiensis* is used and commercialized.

Probably, the least attention to date has been given to insect predators. Their taxonomy and ecology need to be studied before their possible use in integrated pest control can be assessed and ways of multiplying them evaluated (Schaefer and Meisch, 1988).

Preservation of microorganism biodiversity in germplasm and potential of biotechnologies for their utilization

While more than 85 000 accessions of rice are kept in the International Rice Research Institute's rice germplasm bank, other components of the ecosystem preserved in a living state include a few hundred nitrogen-fixing organisms (*Azolla*, cyanobacteria, bacteria) with potential for use as biofertilizers, and a few hundred strains of rice pathogens isolated from the Philippines. There is no collection of invertebrate germplasm.

Improved strains of nitrogen-fixing agents

In view of the rapid progress in genetic engineering, one can speculate on the possibilities of selecting or designing efficient strains of nitrogen-fixing organisms.

Several authors have selected cyanobacteria with high nitrogen-fixing activities. A nitrogenase-depressed *Anabaena* mutant that excretes ammonium ions into the medium was found to provide nitrogen to rice in a nitrogen-free gnotobiotic culture more efficiently than the parent strain (Latorre *et al.*, 1986). But in both cases, strains could not establish themselves in the soil, which is consistent with the observation that no inoculation experiment has yet reported the establishment of foreign cyanobacterial strains in soils (Roger *et al.*, 1991). Significant progress has been made on cyanobacterial genetics, and 'super' nitrogen-fixing cyanobacteria can be selected or probably designed and grown *in vitro*, but the characteristics that will enable them to survive, develop, and fix nitrogen as programmed *in situ* are still unknown.

Azolla collections have been used to screen varieties adapted to specific environments, while some efficient strains have been adopted for practical use. These collections have also been used to achieve the exchanges of cyanobacterial symbionts between species, and for breeding improved *Azolla* hybrids (Roger *et al.*, 1991).

Characterization of inter- and intraspecific diversity of rice pathogens

Sustainable rice ecosystems need to provide, *inter alia*, economic stability to the farmer as demonstrated by stable rice yields. One of the main causes of instability are the sporadic epidemics caused by subpopulations of plant pathogens increasing in frequency relative to the prevailing rice genotype. This results in the so-called 'boom-and-bust' cycles of varietal resistance breakdown (Teng, 1990). Understanding the coevolutionary processes between pathogen populations and rice requires the monitoring of pathogen species at the community level, which was constrained historically by the lack of rapid and accurate methods of characterization. Pathogen identification commonly requires time-consuming axenization of the pathogen to fulfil Koch's postulates. Tools such as polyclonal and monoclonal antibodies have reduced the time needed for identification, but antibodies have only been developed against a few rice pathogen species and this is an area which requires greater effort.

Within the same species, pathogen populations also exhibit much genetic variability on the same rice genotype. The traditional method of studying such variability has been to 'type' the subpopulations using a set of rice differential genotypes representing a range of susceptibilities to the pathogen. With modern molecular markers and selected serological techniques, the tedious and complicated process of race typing has been greatly simplified and made more reliable. Intraspecific diversity of *Xanthomonas campestris* has been recently studied at the DNA level through the detection of restriction fragment length polymorphisms (RFLP) (Raymundo *et al.*, 1990). DNA probes were used to study the partitioning of variability in pathogen populations. Several putative transposable elements were also identified and used to examine the DNA profiles of a collection of strains. The results indicate a particular evolutionary relationship between pathogenic races and rice host resistance. The intraspecific diversity revealed by RFLP typing allows the selection of appropriate tester strains to identify unrecognized races and resistance genes.

Conclusions

The beneficial and detrimental roles of many groups of microorganisms and invertebrates in sustainable rice production have been identified, and sometimes, quantified. However, knowledge on the possible long-term effects associated with the intensification of rice cultivation on these populations is limited. The study of crop intensification effects in long-term experiments should have high priority. In particular, the nitrogen fertility of wetland rice soils depends upon the turnover of a soil microbial biomass representing only a small percentage of total soil nitrogen. Therefore, a general understanding is needed of: (i) the pathways that allow microbial biomass replenishment; and (ii) the long-term effects of crop intensification on microbial and invertebrate populations involved in this replenishment.

Estimates of biodiversity in ricefields are extremely scarce; there are no irrefutable data to demonstrate the generally accepted concept that crop intensification decreases biodiversity in ricefields.

Recent data on arthropods confirm that high biodiversity is not synonymous with stability and with low pressure of insect pests. Increases in pest and vector densities may depend primarily on reduced predator diversity and the resilience and biological attributes of a particular pest. Thus, increasing or preserving diversity *per se* does not necessarily contribute to pest stability, but developing effective trophic linkages might. This approach might also be valid for maintaining soil fertility through microbial management, the optimization of the primary production in floodwater, and the optimization of nutrient recycling by invertebrate populations.

Numerous methods using microorganisms and invertebrates to increase soil fertility (especially through biological nitrogen fixation) and to control pests and diseases have been tested. But the success of these methods is limited and their adoption is almost negligible. This situation will probably remain unchanged as long as the methods designed continue to be based on an extremely restricted knowledge of biodiversity, community structure, and trophic relationships at the ecosystem level.

However, current knowledge shows that there is potential in designing methods that use microbial and invertebrate populations to sustain a management system that reduces and optimizes agrochemical use.

Acknowledgements

P.A. Roger conducts research at the International Rice Research Institute (IRRI) under a scientific agreement between IRRI and ORSTOM (France). Dr J.C. Prot (ORSTOM) and Dr I. Simpson (ODI) are thanked for access to their unpublished results.

References

Ahrens, C., Cramer, H.H., Mogk, M. and Peschel, H. (1982) Economic impact of crop losses. In: *Proceedings of 10th International Congress of Plant Protection.* BCPC, Croydon, pp. 65–73.

Bonman, J.M., Estrada, B.A. and Denton, R.I. (1986) Blast management with upland rice cultivar mixtures. In: *Progress in Upland Rice Research.* International Rice Research Institute, Manila, pp. 375–382.

Bray, F. (1986) *The Rice Economies, Technology and Development in Asian Societies.* Basil Blackwell, Oxford.

Cramer, H.H. (1967) Plant protection and world crop production. *Pflanzenschutz-Nachrichten-Bayer* 20, 1–524.

Denno, R.F. and Roderick, G.K. (1990) Population biology of planthoppers. *Annual Review of Entomology* 35, 489–520.

Dame, D.A., Washino, R.K. and Focks, D.A. (1988) Integrated mosquito vector control in large-scale rice production systems. In: *Vector-borne Disease Control in Humans through Rice Agroecosystem Management.* International Rice Research Institute, Manila, pp. 184–196.

Elazegui, F.A., Soriano, J., Bandong, J., Estorninos, L., Johnson, I., Teng, P.S., Shepard, B.M., Litsinger, J.A., Moody, K. and Hibino, H. (1990) The IRRI integrated pest survey. In: *Proceedings of Workshop on Crop Loss Assessment to*

Improve Pest Management in Rice, October 1987. International Rice Research Institute, Manila, pp. 243–271.

Heckman, C.W. (1979) *Ricefield Ecology in Northeastern Thailand.* [Monographiae Biologicae, No. 34.] W. Junk, The Hague.

Heinrichs, E.A. (1988) Role of insect-resistant varieties in rice IPM systems. In: Teng, P.S. and Heong, K.L. (eds), *Pesticide Management and Integrated Pest Management in Southeast Asia.* International Crop Protection, Beltsville.

Heinrichs, E.A. and Viajante, V.D. (1987) Yield loss in rice caused by the caseworm *Nymphula depunctatis* Guenee (Lepidoptera: Pyralidae). *Journal of Plant Protection in the Tropics* 4, 15–26.

International Rice Research Institute (1989) *IRRI Strategy, 1990–2000 and Beyond.* International Rice Research Institute, Manila.

Ishibashi, N. and Ito, S. (1981) [Effects of herbicide benthiocarb on fauna in paddy field.] *Proceedings of the Association for Plant Protection, Kyushu* 27, 90–93.

Kenmore, P.E. (1989) Development and application of IPM in rice growing in South and Southeast Asia. [Paper presented at 14th Session of the FAO/UNEP Panel of Experts on Integrated Pest Control, Rome, 10–13 October 1989.]

Kenmore, P.E., Heong, K.L. and Putter, C.A.J. (1985) Political, social, and perceptual factors in integrated management programmes. In: Lee, B.S., Loke, W.H. and Heong, K.L. (eds), *Integrated Pest Management in Malaysia.* Malaysian Plant Protection Society, Kuala Lumpur, pp. 47–66.

Kobayashi, T., Noguchi, Y., Hiwada, T., Kanayama, K. and Maruoka, N. (1973) [Studies on the arthropod associations in paddy fields, with particular reference to insecticidal effect on them. Part 1.] *Kontyu* 41, 359–373.

Latorre, C., Lee, J.H., Spiller, H. and Shanmugham, K.T. (1986) Ammonium ion excreting cyanobacterial mutant as a source of nitrogen for growth of rice: a feasibility study. *Biotechnology Letters* 8, 507–512.

Lim, R.P. (1980) Population changes of some aquatic invertebrates in ricefields. In: *Tropical Ecology and Development.* [Proceedings of the 5th International Symposium of Tropical Ecology.] International Society of Tropical Ecology, Kuala Lumpur, pp. 971–980.

Lim, R.P. and Wong, M.C. (1986) The effect of pesticides in the population dynamics and production of Stenocypris major Baird (Ostracoda) in ricefields. *Archiv für Hydrobiologie* 106, 421–427.

Litsinger, J.A. (1984) Assessment of need-based insecticide application for rice. [Paper presented at the 1984 MA-IRRI Technology Transfer Workshop.] Entomology Department, International Rice Research Institute, Manila.

Manwan, I., Sama, S. and Rizvi, S.A. (1987) Management strategy to control rice tungro in Indonesia. In: *Rice Tungro Virus.* [Proceedings of a Workshop held in Maros, AARD, Indonesia.] Pages 92–97.

Mather, T.H. and Trinh Ton That (1984) *Environmental Management for Vector Control in Ricefields.* [FAO Irrigation and Drainage Papers No. 41.] Food and Agriculture Organization, Rome.

Mew, T.W. and Rosales, A.M. (1986) Bacterization of rice plants for control of sheath blight caused by *Rhizoctonia solani. Phytopathology* 76, 1260–1264.

Moon, Y.H. and Kuwatsuka, S. (1984) Properties and conditions of soils causing the dechlorination of the herbicide benthiocarb in flooded soil. *Journal of Pesticide Science* 9, 745–754.

Pathak, P.K. and Dhaliwal, G.S. (1981) *Trends and Strategies for Rice Pest Problems in Tropical Asia.* [International Rice Research Institute, *Research Paper Series,* No. 64.] International Rice Research Institute, Manila.

Petr, T. (1987) Food fish as vector control, and strategies for their use in agriculture. In: *Effects of Agricultural Development on Vector-Borne Diseases.* [Proceedings of the 7th Meeting of the Joint WHO/FAO/UNEP PEEM, 7–12 Sept. 1987.] Food and Agriculture Organization, Rome, pp. 87–92.

Raymundo, A.K., Nelson, R.J., Ardales, E.Y., Baraoidan, M.R. and Mew, T.W. (1990) A simple method for detecting genetic variation in *Xanthomonas campestris* pv. *oryzae* by restriction fragment length polymorphism. *International Rice Research Newsletter* 15, 8–9.

Riessig, W.H., Heinrichs, E.A., Litsinger, J.A., Moody, K., Fiedlr, L., Mew, T.W. and Barrion, A.T. (1986) *Illustrated Guide to Integrated Pest Management in Rice in Tropical Asia.* International Rice Research Institute, Los Banos.

Roger, P.A. (1990) Microbiological aspects of pesticide use in wetland ricefields. [Paper presented at the Workshop 'Environmental and health impacts of pesticide use in rice culture' March 1990.] International Rice Research Institute, Manila.

Roger, P.A. and Bhuiyan, S.I. (1990) Ricefield ecosystem management and its impact on disease vectors. *Water Resources Development* 6, 2–18.

Roger, P.A., Grant, I.F., Reddy, P.M. and Watanabe, I. (1987) The photosynthetic biomass in wetland ricefields and its effect on nitrogen dynamics. In: *Efficiency of Nitrogen Fertilizers for Rice.* International Rice Research Institute, Manila, pp. 43–68.

Roger, P.A. and Kurihara, Y. (1988) Floodwater biology of tropical wetland ricefields. In: *Proceedings of the First International Symposium on Paddy Soil Fertility, 6–13 December 1988.* University of Chiang Mai, Chang Mai, pp. 275–300.

Roger, P.A., Zimmerman, W.J. and Lumpkin, T. (1991) Microbiological management of wetland rice fields. In: Metting, B. (ed.), *Soil Microbial Technologies.* Marcel Dekker, New York (in press).

Rosales, A.M., Nuque, F.L. and Mew, T.W. (1986) Biological control of bakane disease of rice with antagonistic bacteria. *Philippine Phytopathology* 22, 29–35.

Schaefer, C.H. and Meisch, M.V. (1988) Integrated mosquito control in small-scale rice production systems. In: *Vector-borne Disease Control in Humans through Rice Agroecosystem Management.* International Rice Research Institute, Manila, pp. 197–201.

Self, L.S. (1987) Agricultural practices and their bearing on vector-borne diseases in the WHO Western Pacific region. In: *Effect of Agricultural Development on Vector-borne Diseases.* [Proceedings of the 7th Meeting of the Joint WHO/FAO/ UNEP PEEM, 7–12 Sept. 1987.] Food and Agriculture Organization, Rome, pp. 48–56.

Service, M.W. (1977) Mortality of immature stages of species B of the *Anopheles gambiae* complex in Kenya: comparison between ricefields and temporary pools, identification of predators, and effects of insecticidal spraying. *Journal of Medical Entomology* 13, 535–544.

Takahashi, R.M., Miura, T. and Wilder, W.H. (1982) A comparison between the area sampler and two other sampling devices for aquatic fauna in ricefields. *Mosquito News* 42, 211–216.

Takamura, K. and Yasuno, M. (1986) Effects of pesticide application on chironomid larvae and ostracods in ricefields. *Applied Entomology and Zoology* 21, 370–376.

Teng, P.S. (1986) Crop loss appraisal in the tropics. *Journal of Plant Protection in the Tropics* 3, 39–50.

Teng, P.S. (1990) Integrated pest management in rice: an analysis of the status quo with recommendations for action. [Report submitted to the International IPM Task Force.]

Teng, P.S. and Heong, K.L. [eds] (1988) *Pesticide Management and Integrated Pest Management in Southeast Asia*. Consortium for International Crop Protection, College Park, Maryland.

Teng, P.S., Elazegui, F.A., Torres, C.Q. and Nuque, F. (1990) Current knowledge on crop losses in tropical rice. In: *Crop Loss Assessment Rice*. International Rice Research Institute, Manila, pp. 39–54.

Watanabe, I., De Datta, S.K. and Roger, P.A. (1988) Nitrogen cycling in wetland rice soils. In: Wilson, J.R. (ed.), *Advances in Nitrogen Cycling in Agricultural Ecosystems*. CAB International, Wallingford, pp. 239–256.

Way, M.J. (1976) Entomology and the world food situation. *Bulletin of the Entomological Society of America* 22, 127–131.

Discussion

Greenland Could Dr Roger comment on the biodiversity among cyanobacteria in rice paddy fields? Does it relate to the magnitude of biological nitrogen fixation and to the long-term fertility of the paddy soils?

Roger: Our current state of knowledge, based on over 800 papers published on this topic, is that it is not the absence of cyanobacteria which is the cause of the low nitrogen fixation, but that the limiting factors are grazing, a lack of phosphorous, and inhibitory applications of nitrogen fertilizers. Cyanobacteria are now known to be ubiquitous in rice fields and only alleviation of the limiting factors is required. It is important to stress that there is currently no satisfactory taxonomy of cyanobacteria. We lack methods to enable field workers to identify strains. Much published inoculation work does not include records of the strains used and has made it impossible to follow the fate of the inoculated strains. The new biomolecular tools now available could change this.

Eleven

General Discussion: Session II

Session Chairman: J.M. Lynch, *Horticulture Research International, Worthing Road, Littlehampton, West Sussex BN17 6LP, UK*

Session Rapporteur: J.K. Waage, *International Institute of Biological Control, Silwood Park, Buckhurst Road, Ascot, Berkshire SL5 7TA, UK*

Introduction

In this session we concentrated on the environment in relation to biodiversity, and specifically on soil. There has been considerable concern as to the extent to which the soil and its organisms are affected by agricultural practices and pollutants. It is important to emphasize the need to look at organisms in relation to the environment in which they live, as stressed in the vision of *Gaia* proposed by Professor J.E. Lovelock (1982. *Gaia: A New Look at Life on Earth*. Oxford University Press, Oxford). Whether his views are accepted or rejected, the contributions in this session approach the question of biodiversity and sustainability both from the directions of the soil environment and from that of the organisms. The interaction between these provides the basis for soil productivity on the planet.

Discussion

Lynch: In terms of the future directions of work in this area, it is appropriate to mention that the International Union of Biological Sciences (IUBS) has had some initial discussions on the whole question of the ecological importance of biodiversity, including microorganisms, and is considering launching programmes in this area (Di Castri, F. and Younès, T. 1990. Ecosystem function of biological diversity. *Biology International, Special Issue* **22**, 1–20). The CoBIOTECH meeting in Copenhagen in July 1990 strongly supported biodiversity as a direction in which to proceed, and an Expert Meeting convened by DG XII of the Commission of the European Community in June 1990 identified a need to address biodiversity in the plant, animal, and microbial world. The Microbial Strain Data Network

The Biodiversity of Microorganisms and Invertebrates: Its Role in Sustainable Agriculture.
Edited by D.L. Hawksworth. © CAB International 1991.

(MSDN) has been asked by the United Nations Environment Programme (UNEP) to start to establish a database on biodiversity. The Organization for Economic Co-operation and Development (OECD) has also expressed a similar wish. There is a clear message here that there is a major requirement for knowledge on biodiversity.

There are large gaps in our knowledge, and we need to find ways of measuring perturbation in ecosystem biodiversity. This is what most of the concern is about and what the international agencies are seeking. Do we have to go through the long rigorous classical procedures of collecting information or are there any short-cuts? The Shannon Biodiversity Index is being used by some microbial ecologists but has major drawbacks. Baseline information is clearly crucial to quantifying perturbations.

Jeger: With respect to the Shannon Biodiversity Index for quantifying microorganism biodiversity, the problem, especially with fungi, is the difficulty of defining what an individual is. That definition is necessary for calculating most indices of diversity, whether intra- or infraspecific.

Elliott: In approaching biodiversity or sustainable agriculture a systems approach is needed; that is it must be multidisciplinary and probably process orientated. To obtain the database we must be wary of the farming system, for example monoculture or a true rotation, as biocontrol agents in soil respond differently depending on the cropping system. In projecting losses due to pesticides cropping systems are also relevant as these can also destroy beneficial organisms and so the loss projections could be erroneous. The definition of sustainability and biodiversity will have to be a moving target to answer our questions, and must be on a systems approach.

L. Grant: In relation to salinity and biodiversity, the quest is to produce food to meet the demands of a growing world population. Do additions of artificial fertilizers have positive or negative effects on the spectrum of biodiversity in saline soils?

Szabolcs: As soon as such soils start to be utilized, water and chemicals are added. The biota is consequently changed radically from that of a desert soil to a cultivated soil, but it is difficult to categorize this as positive or negative or to state the extent. The particular situations must be studied on the spot to provide such answers.

P. Williams: With respect to the measurement of biodiversity, the Biodiversity Programme of The Natural History Museum includes a project to measure taxonomic diversity, particularly diversity by ancestry. Some of the early results from this programme can now be demonstrated.

Swaminathan: Some plants have the capacity to either tolerate or extract salts from the soil and have been used in overall reclamation strategies. *Anemone mexicana*, for example, has given good results in alkaline soils in India. In coastal areas mangroves can also tolerate seawater intrusions. I feel we should breed plants which are efficient in the extraction of salts from soil.

Szabolcs: Where more radical reclamation methods cannot be used, employing plants is a good and traditional method. There are many successful species, as for example using *Puccinellia limosa* which is cheap and simple; in some cases even *Acacia* and mangroves have also been used. The approach should be extended in the future, particularly as it lacks the economic and other hazards that can be associated with radical mechanical and chemical reclamation programmes.

Perfect: It seems that to talk of conserving microorganisms is a forlorn quest. Microorganisms appear to be able to adapt rapidly to changing situations. How can we establish a databank on their biodiversity when what we would really be doing is to reflect their ability to respond to changing conditions. We need to exploit and harness the capabilities of microorganisms, but is it realistic to talk about their biodiversity as a phenomenon to capture and exploit?

Bull: I am increasingly pessimistic about taking a taxonomic approach, at least with regard to bacteria. An alternative pragmatic, but less satisfying non-systematized approach might be based on functional properties selected for specific purposes by agronomists, biotechnologists, and other applied scientists. Taxonomic analysis would be a subsequent and complementary action. The extent of microbial biodiversity probably mirrors that at the macro-level, especially in view of the intimate associations between plants and other organisms. The first priority is thus to conserve habitats.

Lynch: I agree that bringing in samples from different habitats and searching for particular properties is an important approach and one used in my laboratories. Having found organisms of interest, it is then that the services of CAB International become vital when the critical taxonomic information is required.

Holloway: With respect to beneficial earthworms in soils, while not diverse in themselves it would be of interest to know the extent to which they are dependent on the diversity of other soil organisms. Conversely, is there any evidence that 'bad' soil organisms are encouraged by poor husbandry? There is a danger of failing to appreciate the interdependence of life, and I suggest that we need to identify the effective linkages referred to by Dr Roger (Chapter 10).

Rothschild: There appears to be a need for experimental systematic work on biodiversity, experimental manipulation of the environment to simulate different farming practices, etc., and then monitor their effects on biodiversity. This experimental approach might provide opportunities for different groups to collectively examine key issues in a comparable way. A network drawing together researchers in collaborative projects might be envisaged. This could, for example, provide indications of critical areas in terms of size for conserving sources of biodiversity. In the case of saline environments, would it be necessary to conserve salty areas as a source of salt-tolerant organisms? Were there to be a widespread amelioration of saline areas, such sources may be greatly reduced.

Szabolcs: I agree that our knowledge of the systematics of soil organisms, not only of saline soils, is far from complete. There is also a systematics of soils that is quite separate. Sooner or later this deep gap must be bridged, and biological data taken more into account in pedological classifications. This will not be a rapid process but step by step it is necessary to proceed along this route, especially if we are to be able to manage soils in relation to their biotas. The development of salt-affected soils amounts to only about 5% of the total area of such soils on earth, and so conservation is not at present a concern. The most typical places should nevertheless be protected.

Greenland: Soils affected by salinity or severe erosion are often largely devoid of life. The addition of organic material, and reinoculation, is far from adequate for the restoration of a soil to its former state. The biodiversity which accompanies an active soil appears to be necessary for most soils if they are to be productive. We need to establish methods by which biodiversity can be assessed in terms of what is necessary for an active and productive soil; it must include those organisms that ensure and active turnover of organic material. Knowledge of the taxonomy of microorganisms is important here; it would have saved considerable time in the work carried out on cyanobacteria in rice fields (p. 136). An example of a problem is the specific toxicity of chloride ions to most nitrifying microorganisms. This means that nitrogen once mineralized to ammonium will be lost as ammonia gas rather than be nitrified; if a chloride-tolerant nitrifier exists it is important that it is not lost.

Lynch: Inoculating soils is extremely difficult. In experiments on volcanic ash on Mt St Helens (WA, USA), it was found that if microorganisms were inoculated they would do nothing as they had no substrate base. Inoculations would need to be on an appropriate plant base in order to regenerate soil structures.

Hulse: There is a severe incidence of human toxicity caused by package labels that cannot be understood by farmers and their families. The International Rice Research Institute (IRRI) has considerable experience in reaching farmers through its farming systems and training programme. That experience could be of particular help in ensuring that recommendations form this and other scientific meetings are translated into communication media that can be understood and used.

Simpson: I can endorse from my own investigations the problems caused by inappropriately or inadequately labelled packaging. Communications between scientific organizations and farming communities are clearly crucial.

F. Baker: The Technical Advisory Committee (TAC) of the Consultative Group on International Agricultural Research (CGIAR) has proposed the following definition of the concept of sustainable agriculture: 'it should involve the successful management of resources for agriculture to satisfy changing human needs while maintaining or enhancing the quality of the environment and conserving natural resources'. Should the Workshop be more concerned with the question of the sustainable management of underexploited natural resources, such as the larger algae or locust swarms, in order to try to alleviate the effects of increasing population pressures?

Session Chairman's conclusions

The suggestions and targets emerging from many of the papers in this Session were consistent with those from many international agencies concerned with the environment. For example, at a recent experts' meeting of the OECD Research Project on Biological Resource Management, the topic of modification of plant/soil/microbial interactions to reduce inputs in farming systems identified the study of soil quality and biodiversity as one of the areas needing support in terms of workshops and fellowships. It was considered that heavy metals, pollutants, fertilizers, and salinity could all influence ecological behaviour. Methodology should be developed to determine diversity indexes in relation to both functional and genetic diversity. Predators probably play a crucial role in this. Those conclusions seem in no way inconsistent with the findings of the present Workshop.

In soils, one of the major factors to consider is perturbation, and implicit in this is that the baseline ecology is well-understood. This is seldom the situation and there is a need for further inputs at the functional and systematic levels.

Functional analysis should be considered in terms of benefits. A proportion of the biota will carry out beneficial functions, such as providing nutrients to plants, whereas other such as pathogens are harmful. A reduction in biodiversity is likely to be useful if *only* the harmful components are eliminated.

Some of the beneficial features of the biota could be enhanced by their improvement, phenotypically or genotypically. Also, modification of the plant by conventional breeding or genetic manipulation is likely to modify the rhizosphere and soil biota (Lynch, J.M., ed. 1990. *The Rhizosphere*. John Wiley, Chichester).

Biodiversity varies in different climatic regions and land use also has an influence. There is a need for comparison between such systems as tropical rain forest, temperate arable and arctic tundra ecosystems.

To achieve such targets of study, there is an urgent need to develop new methods to determine the role of biomass components and the nature of individuals inter- and intraspecifically. Experimentalists, taxonomists, and mathematicians, need to work on an interdisciplinary basis.

At this stage biodiversity is difficult to determine mathematically (May, R.M., 1976. *Theoretical Ecology, Principles and Applications.* Blackwell Scientific Publications, Oxford) and whereas we should strive to produce definitions, there is an urgent need for experimentalists to report results, for example of numbers *versus* individuals, so that theoretical ecologists/ mathematicians can develop concepts. Such a structured approach would avoid the tendency to allow the analysis of biodiversity to develop by serendipity.

III
Importance of Biodiversity to Pest Occurrence and Management

Twelve

Biodiversity and Tropical Pest Management

T.J. Perfect, *Natural Resources Institute, Central Avenue, Chatham Maritime, Chatham, Kent ME4 4TB, UK*

ABSTRACT: The term biodiversity has been coined to describe variation in biological systems; it is convenient, but so all-embracing as to demand dissection in the context of a presentation such as this. It has long been an article of faith among ecologists that diversity equates with stability. With certain exceptions this is broadly true; floristic diversity within the habitat generates the architecture and range of food sources to support faunal diversity and create ecological niches bounded by nutritional or behavioural specialization.

Introduction

Diversity creates a buffering capacity in the sense that perturbations have little effect on the functionality of the system through the ability of organisms to exploit adjacent niches made accessible through changed conditions. This may lead to a reduction in biodiversity without effects on ecosystem function or productivity. Such effects are of prime concern in a consideration of pest management; indeed, modern approches to pest management exploit and promote biodiversity within agricultural systems. This contribution aims to provide an overview of the key concepts, particularly in relation to the tropics.

Habitat diversity

Any consideration of agro-ecosystems must be set within the framework of

The Biodiversity of Microorganisms and Invertebrates: Its Role in Sustainable Agriculture.
Edited by D.L. Hawksworth. © CAB International 1991.

the physical parameters that constrain their productivity. The variables of soil type and climatic characteristics are not for treatment here, but most systems capable of supporting forest climax vegetation have diverse agricultural potentials, providing soil fertility can be maintained; rangeland and savanna are more suited to the production of legumes and cereals. Improved water management can extend production in arid and semi-arid systems and increase it, notably through irrigated rice, in the humid tropics.

The indigenous farming systems of the tropics, particularly in the humid and sub-humid zones, are characterized by their diversity and often some form of shifting cultivation that maintains a reservoir of the natural flora and fauna. This has arisen through the needs of small-scale subsistence farmers in the face of a series of constraints, social, economic and logistic, to promote local food security and self-sufficiency. Improved communications, the development of an infrastructure supporting a market economy, population increase and urbanization with the ensuing pressure on land, have all contributed to a shift away from traditional farming practices. The trend has been towards larger areas under production, fewer crops, and higher agricultural inputs.

Restricting the production base finds its ultimate expression in crop monoculture; an extreme reduction in the interspecific biodiversity of the cropping system. This has a profound effect on the dynamics of the pest complex. Pest species are particularly well-adapted to exploit the patchiness in distribution of their host-plants, both in terms of dispersal and host location and the rapid build-up of populations after colonization. Increasing patch size and homogeneity both favour specialist pests and population development frequently 'escapes' from the natural regulating capacity of the system.

Plant biodiversity

This effect is compounded by a narrowing of the genetic base of the crops themselves arising as a consequence of breeding programmes with a primary emphasis on yield. Most traditional cultivars are low yielding, but robust in terms of their response to pests; the characters conferring this robustness, or 'cryptic resistance', are often lost during selection designed to improve yield. Thus, intraspecific variability is reduced. This triggers a cycle of events not unlike the well-known 'pesticide treadmill' where pest species, themselves genetically highly plastic, develop the capacity to overcome widely deployed varieties with monogenic resistance and the result is often disastrous crop loss. Wheat rusts and the rice brown planthopper are good examples of this phenomenon.

The biodiversity existing in unimproved cultivars of crop plant and their wild relatives is a resource of the utmost importance in deriving new sources of resistance. Most crop breeding programmes are working towards strengthening genetic diversity within crops through combining different resistance characteristics within a single cultivar or through mixed plantings of phenotypically similar varieties carrying different resistance genes. It is hoped this will lead to what has become known as 'durable' resistance.

Faunal diversity

Earlier reference was made to a reduction in faunal diversity within cropping systems resulting from a shift towards monoculture (Holloway and Stork, Chapter 5). The net effect is to favour particular well-adapted species, some of which therefore assume pest status. For many annual crops these are invasive species. The discontinuities created by large scale, intensive crop production practices reduce the background level of generalist predators, and of more specific parasitoids, and give rise to a further condition favouring pest population development.

Pesticides are another factor to be taken into account in this regard. The ideal pest-specificity and precisely timed and targeted application technology are rarely available; even if they are, logistic constraints commonly prevent their realization in the field. Pesticides frequently further reduce biodiversity through differential toxicity or access to different components of the faunal complex. Pests are more robust than natural enemies for a number of reasons and the effect is often to deregulate population growth. This is not common early in the life of a chemical pesticide, but there are many examples of the phenomenon associated with the development of pesticide resistance. There are analogies, as implied above, with the management of crop varieties.

Diversity in pest management

Biodiversity within pest species has a further importance in addition to that of determining pest status. It can also specifically affect the choice of control technology. This goes beyond the obvious instances of varietal and pesticide resistance to such techniques as the use of pheromones for pest monitoring or control through mating disruption. The development of host-plant races in pathogens and nematodes, and of biotypes in insects, are special instances of biodiversity for which a discriminant capability is critical in deciding on the management strategy to adopt. Modern techniques are beginning to provide the methodology to improve the basis for making such decisions, though conceptual difficulties remain in proper deployment of the resulting information.

Conclusion

It would be unreasonable to stretch even such an elastic term as biodiversity to encompass the use of varied approaches in the formulation of pest management strategies. However, this is such a key concept in sustainable pest management that it demands some mention. All management interventions have other than their intended impacts on the faunal and floral complex associated with agro-ecosystems. Some are outlined above. The impacts vary according to the nature of the intervention, and it makes good sense to avoid over-reliance on a single tactic as a means of retaining robustness and variability within the system.

Discussion

See under Chapter 13 (pp. 162–3).

Thirteen

Biodiversity as a Resource for Biological Control

J.K. Waage, *International Institute of Biological Control, Silwood Park, Buckhurst Road, Ascot, Berkshire SL5 7TA, UK*

ABSTRACT: Biological control makes use of indigenous and introduced natural enemies to provide sustainable control of pest populations. The pool of plant feeding organisms, predators, parasites, and pathogens from which we draw biocontrol agents constitute over 50% of species on Earth. Hence, the subjects of biodiversity and biocontrol are linked. At any site, the specificity of natural enemies will make them at least as diverse as the pests they attack. Diversity is probably depressed in agro-ecosystems due to cropping practices. For instance, some important natural enemies such as fungi and ants are excluded by cultivation. Natural vegetation around crops is important to conserving natural enemy diversity, and possibly more so in tropical than temperate systems. Presently, natural enemy diversity is exploited on a global scale in the search for new biological pesticides and exotic natural enemies to control introduced pests. Demand for both will increase, and there is a need to conserve source areas for these natural enemies, presently threatened by habitat destruction. More generally, there is a need to characterize little known natural enemy faunas, particularly in tropical systems.

Introduction

'Biological control' has enjoyed many different definitions. In this contribution it is taken to mean the use of living organisms as pest control agents. These living organisms, or natural enemies, constitute an important resource for crop protection, because they kill pests and, in the process, reproduce so as to perpetuate the control which they exert.

The Biodiversity of Microorganisms and Invertebrates: Its Role in Sustainable Agriculture.
Edited by D.L. Hawksworth. © CAB International 1991.

Biological control seeks, first and foremost, to conserve natural enemy species in the crop environment. When this is not sufficient, natural enemies may be regularly added to the crop system, or new species may be introduced and established from other regions to reduce pest populations. Together, these different methods of biological control constitute a sustainable and environmentally safe component of a modern pest management, and one which is seen increasingly as a desirable alternative to dependence on chemical pesticides.

Natural enemies of insect pests include their predators, parasites and pathogens. The natural enemies of weeds include plant-feeding insects and pathogens, while plant diseases themselves have pathogens which may control their growth, as may some antagonistic and competing microorganisms, which are also sometimes considered agents of biological control.

Clearly, the natural enemies which we use in biological control are a varied assemblage of organisms, about which it is difficult to generalize. However, one consistent feature of taxa containing natural enemies is their remarkable taxonomic diversity. The parasitic life-style, which many natural enemies exhibit, typically leads over evolutionary time to a great degree of specialization, which in turn creates a great number of distinct species (Price, 1980). This is nowhere more clear than with insect natural enemies. Species of plant-feeding insects, which include biological control agents for weeds as well as crop pests, together with their insect predators and parasites, comprise over 50% of the species of life on our earth (Strong *et al.*, 1984; J. Lawton, personal communication). Among these insect natural enemies, the parasitic wasps or parasitoids, which attack insect pests, have been estimated to comprise close to 10% of all species of animal life (Askew, 1971). Statistics of insect diversity, presented by Hawksworth and Mound (Chapter 3), may be used to transform these estimates into numbers of species. The result is impressive, both on a grand scale and when we focus in on particular families and genera: some single genera of parasitoids, for instance, are thought to contain as many as 1 000 species (Gauld and Mound, 1982).

As natural enemies make up such a large component of life on earth, processes which influence global biodiversity will no doubt influence their diversity. This will apply in both natural and agricultural systems, and in the latter may affect their use in biological control. On this basis, the subjects of biodiversity and biological control are linked. In this contribution, I will explore this linkage from several perspectives. First, I will identify the ecological processes which influence the biodiversity of natural enemies, and its distribution around the world. Then, I will consider the diversity of natural enemies in crop systems, and factors which limit it. Finally, I will consider how the global diversity of natural enemies may be utilized in the development of effective biological pest control, and how this valuable resource might be protected for use by future generations.

The source of natural enemy diversity

Ecologists today see the trophic structure of ecosystems as a major determinant of their biodiversity. Simple trophic chains, from primary producers (plants) to primary consumers (plant-feeding organisms) to secondary consumers (predators, parasites, pathogens) are usually characterized by increasing species diversity at each level. This diversity, in turn, appears to depend largely on the amount of resource available for exploitation by each successive trophic level. Thus, recent studies have shown that the diversity of species of primary consumers is a fairly consistent power function of the productivity of primary producers (McNaughton *et al.*, 1989).

This relationship applies not only to natural but to agricultural ecosystems, as revealed in a series of studies on the relationship between the area covered by a crop and the number of insect pest species associated with it. Data from crops as diverse as cocoa, coffee, citrus, sugarcane, and rice reveal a consistent exponential function between the area of cultivation or level of production and the number of insect pest species recorded from the crop in that region (Strong *et al.*, 1984). Surprisingly, this relationship appears to be independent of the area of origin of the crop: while cocoa is a crop native to tropical America, the greatest number of cocoa pest species are found on cocoa in Ghana, the country with the highest level of cocoa production (Strong, 1974).

A number of plausible explanations exist for this species–area relationship, which has also been found in plant parasitic fungi as well as insect pests (Strong and Levin, 1975). On the one hand, larger amounts of resource may provide more opportunities for colonization and support more species because of reduced competition, as embodied in the equilibrium theory of island biogeography (MacArthur and Wilson, 1967). It is also possible that larger areas of resource are distributed over a more diverse environment, providing more variations in habitat which affords more discrete niches for colonization by different species.

This latter hypothesis, that a more complex environment provides more space for different, specialized species, is borne out even on a 'micro' level. Plants with a complex architecture, for example tree crops like coffee or citrus, support many more species of pests than crops with a simpler architecture, such as cereals.

Thus, we have a reasonable understanding of the patterns, and some of the causes, of biodiversity at the level of the primary consumer, which has some relevance to the biological control of weeds, where we use plant-feeding insects as biocontrol agents (Lawton and Schroeder, 1978). Unfortunately, when we move up to secondary consumers, where most of our natural enemies reside, our level of understanding falls sharply.

Perhaps the most confident hypothesis which we can make about the diversity of natural enemies at this level is that it will be at least as great as that of their prey, simply by virtue of their high degree of specificity to particular host species. For example, Price (1980) has analysed host range data from systematic studies on some groups of parasitic wasps: in one study of 214 species of Braconidae attacking agromyzid flies on British Umbelliferae, 60% of species attacked only one plant-feeding insect host. In another study of 514 species of North American Ichneumonidae, 53% of the species were recorded from only one host species.

The diversity associated with this high degree of parasitoid host specificity is compounded by the fact that most insects support parasitic wasp species which attack different life stages (e.g. egg parasitoids, larval parasitoids, pupal parasitoids). This further increases the size of parasitoid complexes on particular insect host species.

Specificity may be even higher among the pathogens of insects and plants. Insect baculoviruses attacking the Lepidoptera were once thought to be highly specific and, although this view may now need to be modified somewhat (Doyle *et al.*, 1990), a high degree of specificity remains characteristic. Insect parasitic fungi may show similar patterns: it is suspected that the family Laboulbeniaceae, which is largely restricted to beetles, may contain species which themselves are restricted to a single host. With well over a quarter of a million known species of beetles, this could be a great many fungal species indeed, as discussed by Hawksworth (1991).

With respect to plant diseases, the rust fungi, which have been most widely used in biological control, comprise over 6000 species. Rusts occur on more different kinds of vascular plants than any other pathogen group, and are commonly specific at the genus or species level (Hart, 1988): hence their real diversity may in some cases be much more than recorded at present.

Beyond the diversity generated by the extreme specificity of many natural enemies, and hence indirectly by the diversity of their prey, we can predict that natural enemies will be subject to the same ecological processes outlined above for primary consumers. That is, their diversity should increase with the abundance and distribution of the prey species. For parasitoids, it is known that the complex of species attacking a host over its entire geographical range is likely to be larger than that attacking it in any one area (Askew and Shaw, 1986; Hawkins, 1988), but this is not a proper test of the species–area relationship and may also represent a sampling artefact – larger samples turn up more species. There is evidence from the biological control literature (e.g. Bess *et al.*, 1961; DeBach *et al.*, 1971), and from ecological research and theory (Hassell and Waage, 1984), that interspecific competition may limit the size of parasitoid complexes attacking particular host species. Hence, the equilibrium theory of island biogeography may be applicable here, with greater resources affording greater coexistence.

Some notable global patterns in natural enemy diversity have been found. For some parasitoid groups, a number of studies have shown that diversity is actually lower in tropical than in temperate regions (Askew and Shaw, 1986; Noyes, 1989; Hawkins, 1990). The popular hypothesis for this consistent pattern is that the high diversity but low local density of tropical herbivorous insects (in part due to intense predation) makes the support of highly specific parasitoid species difficult. Tropical parasitoids appear to be more general in their habits, a potentially important character discussed further below (p. 154).

By contrast, ants, which are themselves generalist predators, are not only extremely important in the tropics, but far more diverse there (Kuznezov, 1957). Their fungal pathogens are also equally diverse, and this reflects a general trend in entomopathogenic fungi to be more variable in tropical habitats (Evans, 1982).

Natural enemy diversity in agro-ecosystems

Agricultural systems are generally simpler than the natural habitats from which they are developed. They have fewer plant species, fewer primary consumers, and, in general, fewer natural enemy species. For particular pests, natural enemy diversity may be lower in crops than in natural habitats. Thus, in extensive surveys for the parasitoids of *Helicoverpa armigera* in the Mediterranean region, Carl (1989) found only two of the 12 known species of parasitoids attacking the pest in crop monocultures, the others presumably being restricted to natural habitats. Contrasting patterns, however, can occur: the moth pest, *Spodoptera praefica*, is attacked by more parasitoid species in alfalfa fields than in natural grassland (Miller, 1980).

Where natural enemy diversity is lower in crops, the explanation can often be attributed to the effects of cropping practices on natural enemy survival. Short-cycle crops are often difficult places for the persistence of natural enemy species, because the pest disappears with the crop when harvested. It is not surprising, therefore, that attempts to establish new, exotic natural enemies into such agro-ecosystems has been less successful than in more permanent crops, such as orchards (Hall *et al.*, 1980; Greathead and Waage, 1983).

Certain key groups of natural enemies are excluded by cultivation in arable crops. In the tropics, insect pathogenic fungi can be important control agents of insect pests. The most diverse genus of these entomopathogens in primary forest systems, *Cordyceps*, produces fruiting bodies from which spores are slowly released. These structures are disrupted by cultivation, and in this way many species of potential natural enemies are excluded from tropical crop environments (Samson *et al.*, 1988).

Similar problems can be encountered with insect natural enemies. Ants are ubiquitous and efficient predators which establish easily in plantation crops, where they build nests in the ground or in trees. In short-cycle field crops, however, cultivation usually disrupts ant nests and makes it difficult for them to establish. The potential contribution of ants in such agroecosystems has recently been revealed in a study by the International Institute of Biological Control of biological control in smallholder crops in Kenya. Here, where plots were small, natural vegetation abundant and tillage limited, *Iridomyrmex* ants were found to be the single most important natural enemies of the moth pest, *Helicoverpa armigera*, on a range of crops (H. van den Berg, unpublished data).

Many other cropping practices can, in principle, reduce local natural enemy diversity, not the least of which is the widespread use of pesticides. The action of pesticides on natural enemy diversity is, fortunately, usually only temporary, but long-term studies of pesticide use have revealed that persistent spraying can eliminate species of less mobile natural enemies from the local crop environment (Burn, 1989).

Because crop environments are often inhospitable, many natural enemies persist in the natural habitats around, invading crops continuously, or establishing there for a season, until the crop is taken from the ground. Many studies have identified the importance of natural vegetation around crops to the maintenance of natural enemies (van Emden, 1990). In some cases, natural vegetation provides shelter, in others, an alternative prey between crops, or other food source such as flowers. Polyculture has been suggested as a means to increase vegetational, and hence natural enemy diversity, within crops (Letourneau, 1987).

If a substantial proportion of the natural enemy 'reservoir' persists in natural habitats near crops, the preservation of these habitats is essential to biological control. Ecological theories discussed above are relevant here: the number of species which crop surrounds can maintain will probably be a function of their area and extent. When crops are islands in a sea of natural habitat, the natural enemy reservoir will be large. But as natural habitats become smaller and smaller islands in a sea of cropland, the number of species which those islands can support will decrease, and important natural enemies may become, at least locally, extinct. This applies to both temperate and tropical agriculture, but our present limited knowledge suggests that the situation in the tropics may be more fragile.

Some natural enemy groups, such as insect parasitoids, appear to be less diverse and more generalized in the tropics. Generalist predators, like ants and wasps, are particularly important in tropical systems. All of these generalist natural enemies may depend more on non-pest species in natural habitats for their persistence. If these habitats are reduced, the diversity of non-pest species will be affected as will that of these natural enemies.

It is appealing to think that the tropics, with their benign climate and high productivity, would be particularly resilient to agricultural perturbations of the above kind, and therefore that natural enemy population dynamics would be more stable than in temperate regions. As we learn more about the ecology of natural enemies in the tropics, however, we discover that their populations can be highly sensitive to disruption. For instance, tropical pest populations usually overlap in time, and many tropical natural enemies depend on the constant availability of all life stages of the pest. Disruption of this pattern, by seasonal cropping or pesticide use, may lead to the local absence of certain natural enemy species. When overlapping generations of pests become synchronized by pesticide applications, natural enemies with generation times shorter than the pest cannot easily persist, and may drop out. This explanation has been suggested for the pesticide-induced outbreaks of moth pests on plantation tree crops in South-East Asia (Godfray *et al.*, 1987; Perera *et al.*, 1988).

In summary, it is important to observe that farmers and pest management specialists are not so much interested in the diversity of natural enemies in crops, but in the abundance of those key species which affect pest populations. Hence, maintaining diversity is not, in itself, the key to effective biological control, but insofar as conditions ensuring an abundance of natural enemies may be similar to those ensuring a richness of species, diversity can be linked to the success of biological control. It therefore seems reasonable to suggest that agro-ecosystems have lower natural enemy diversity than natural ecosystems, and that this diversity may be lowered further by certain cropping practices and the reduction of natural habitats relative to cropland. Finally, tropical systems may be more sensitive to such changes than temperate ones.

Making use of biodiversity

Biopesticide development

Where the natural enemies present in the agro-ecosystem are insufficient for effective pest control, an alternative method of biological control, the augmentation of natural enemies, is a possibility. A range of natural enemies can be mass produced and released into crops. For many decades, for instance, tiny parasitoids of eggs, *Trichogramma* species, have been mass produced and released in crops at rates of up to 500 000 per hectare to control moth pests. Today *Trichogramma* gives good biological control of several tropical and temperate pests, particularly stem-boring moths in maize and sugarcane.

However, most development of augmentative methods is now focused on the pathogens of insects, weeds, and plant diseases. It is not uncommon to find highly virulent pathogens attacking pests, but their limited dispersal and survival abilities frequently cause them to have little impact. Mass produced and applied as a pesticide, however, these pathogens can be effective over large areas, without the undesirable environmental consequences associated with broad-spectrum chemical compounds.

Many pathogens can be produced cheaply on artificial media and thereby bulked for formulation and application as biopesticides. They may originate from the very field where the target pest is to be controlled, but commercial formulations are more likely to originate from one site and to be used widely elsewhere. Because of this, it is of particular interest that the pathogen selected for formulation is the most virulent species or strain available.

In order to find highly effective pathogen strains and species, extensive surveys of naturally occurring pathogens are undertaken, collecting and testing isolates from all over the world against key target pests. Sometimes it has been found that the most virulent strains are not those which come from the target itself, but from other pest species. Thus, in a study of fungal pathogens for control of the stem boring moth *Diatraea saccharalis*, the most virulent strain of the fungus, *Metarhizium anisopliae*, was not one from this species but one from the sucking bug *Zulia entreriana* (Alves *et al.*, 1984).

The search for new species and strains leads not only to the improvement of biopesticides but to the development of entirely new ones for new pest groups. This is illustrated by the recent history of the most popular bioinsecticide, a formulation of *Bacillus thuringiensis*. *B. thuringiensis* has been marketed for over 20 years for the control of moth pests. The growing popularity of this pesticide resides in its safety to indigenous natural enemies, humans, and the environment. But this safety resides, in turn, in its specificity to moths.

The plant protection industry has been interested in finding *B. thuringiensis* strains which might be effective against other pest groups. This aim was realized in 1976, when serotype H-14, a strain effective on mosquitoes and blackflies, was found. This was later called *B. thuringiensis* var. *israelensis* (World Health Organization, 1982). Formulations of this pathogen now underpin programmes for the control of the vector of river blindness in West Africa. In 1982, after considerable additional exploration and research, another variety, var. *tenebrionis*, was discovered (Krieg *et al.*, 1983) which attacks beetles, and holds much promise for the control of storage pests.

The rich, global biodiversity of bacterial, fungal, and viral pathogens of insects and other pests remains a little exploited resource for biological control. Commercial interests have not limited the search for new pathogens to agricultural systems. Tens of thousands of pathogen isolates are being collected annually from tropical rain forests and also other natural habitats.

Admittedly, the aim of this exploration is not simply to find biopesticides. Indeed industry may be more interested in isolating the active new metabolites from these pathogens for development as chemical pesticides or pharmaceuticals. But such search can no doubt also lead to the discovery of new biopesticides as well. Some of these will become new pest management products, destined for sale in the developed world which can afford them. Ironically, farmers in the developing world may never have the resources to benefit from pathogens from their own agricultural and natural habitats. Thus, could biological control agents follow the well-worn route of so many other tropical natural resources, never to be fully enjoyed by the countries which provided them?

However, there is an alternative scenario in this case. The production of pathogens for pest control does not require complex technology, and it is often suited to the labour intensive production systems characteristic of tropical agriculture (Prior, 1989). It is possible, therefore, that tropical countries could themselves exploit their rich diversity of naturally occurring biological control agents through local commercial enterprise, at much reduced cost. Precedents already exist. In Brazil, an indigenous strain of *Metarhizium anisopliae* is produced commercially for the control of the sugar cane froghopper *Aeneolamia varia* over thousands of hectares (Prior, 1989). Elsewhere in the same country, an insect virus is produced to control a moth pest on soyabean, with current application over 500 000 ha year^{-1} (Moscardi, 1989). About 10% of this application originates not from the commercial formulations but from farmers themselves, who collect, macerate and apply suspensions of diseased caterpillars back onto their crops (D.L. Gazzoni, personal communication). The added benefit of this approach is that safe, specific controls may be found for serious but local tropical pests which, for market reasons, may never attract the sustained interest of multinational agribusiness. Helping developing countries to develop new biopesticides for local use is a major activity of the International Institute of Biological Control, which presently is undertaking surveys for pathogens to control various tropical pests, such as the desert locust (Prior and Greathead, 1989).

Classical biological control

The global exploration for valuable biological control agents has a precedent which pre-dates by almost a century the current interest in biopesticides. This is the method called 'classical' biological control, by which new, exotic pests are controlled by the introduction of specific natural enemies from their areas of origin.

Classical biological control addresses the continuing and increasing problem posed by the invasion of pests into new regions of the world. These exotic pests, released from their specific natural enemies and poorly controlled by indigenous species, frequently reach outbreak proportions. The appearance of the neotropical larger grain borer (*Prostephanus truncatus*) in Africa, the Eurasian cotton whitefly (*Bemisia tabaci*) in tropical America, the Indian mango mealybug (*Rastrococcus invadens*) in West Africa, the African coffee berry borer (*Hypothenemus hampei*) in Central

America, and the neotropical water hyacinth (*Eichhornia crassipes*) in East Africa, represent just a few of the new exotic pest problems of the past few decades.

The introduction of specific natural enemies from the area of origin of a pest has proved very effective in the control of many exotic pests. Over the past hundred years, over 500 insect natural enemies have been established for control of exotic insect pests, with a success rate of about 40%, while over 100 insect species have been established for the control of exotic weeds, with a success rate of about 30% (Waage and Greathead, 1988).

It is significant that in none of these programmes has there been any evidence of substantial harm to natural or agricultural ecosystems as a result of the introduced natural enemies attacking other, desirable, species. Further, the return on investment in such programmes has been considerable (Tisdell, 1990). In a programme against the exotic mealybug *Rastrococcus invadens*, run by the Togolese Service de la Protection Végétaux with the assistance of the International Institute of Biological Control, the cost of introducing the successful parasitoid (*Gyranusoidea tebygi*), was approximately US$172 000, whereas annual benefits in increased fruit production may be more than 10 times that figure (Voegele *et al.*, 1991). Much of the benefit of classical biological control can be attributed to the fact that, once successful, the control perpetuates itself, at no cost to the farmer.

Classical biological control draws heavily on the biodiversity of natural enemies around the world. Very often, an exotic pest is not a pest in its area of origin, indeed it is sometimes completely unknown, and its natural enemies unstudied. For this reason, classical biological control is often limited by our understanding of the indigenous natural enemy faunas of different countries, which is today very poor. Biosystematics pays a crucial role here, indeed biological control has often failed through errors in establishing the true identity of new pests and natural enemies.

The success of this method is also increasingly constrained by reduced access to the original populations of the pest, on which any possible control agents will be found. Some sources of classical biological control agents are being threatened by habitat destruction. Thus, in eastern Europe, a source area for insect natural enemies of innocuous wild flowers which have become devastating weeds in North American grasslands, intensified herbicide application on road verges has greatly reduced natural populations of these plants and, hence, the local abundance and diversity of potential biological control agents (D. Schroeder, personal communication).

Other sources are extremely local and therefore at risk. Thus the suspected original range of *Ageratina riparia*, a serious exotic weed of pastureland throughout the tropical world, is a single remote valley in Mexico, where it was first described in the mid-1800s, and where it persists today. From its leaves, scientists of the International Institute of Biological Control have collected a white smut fungus which has proven successful as a classical biological control agent for this weed in Hawaii (Baretto and Evans, 1988).

What other obscure and rare indigenous plants, insects and pathogens will become the devastating exotic pests of tomorrow? How certain are we that they and their specific natural enemies will be preserved for future classical biological control programmes? What strategy should we adopt for this preservation? As we cannot pin-point the species we will need, the only effective strategy (as with the preservation of resources for plant breeding or pharmaceutical products) is to preserve entire natural ecosystems, particularly in tropical regions where many of these problems originate.

In a recent magazine advertisement by the Tropical Forest Project of the World Resources Institute, a number of organisms from the neotropical rain forests are illustrated as justifications for the preservation of this important habitat. Predictably, they included potential crop plants, wild varieties of existing crops, sources of drugs, attractive birds and, somewhat incongruously, a tiny beetle of the genus *Phyllotreta* (correctly *Agasicles hydrophila*). This beetle is a herbivore of the water-weed *Alternanthera phylloxeroides*, a serious exotic pest of North American waterways. It is therefore, a biological control agent, and its status in such a distinguished list of rain forest resources indicates the acknowledgement by conservation organizations and agricultural scientists alike, of the need to preserve the organisms which, at some future date, will allow us to restore the balance of nature in other lands.

Conclusion

In this contribution, I have endeavoured to show that maintaining the diversity of natural enemy species in natural as well as agricultural systems is an essential prerequisite to the successful future use of biological control in pest management. This requirement applies equally to the conservation of local natural enemies and the utilization of exotic species for local pest control.

In tropical systems, despite their great biodiversity, natural enemies may be particularly susceptible to local extinction as a result of habitat destruction and unfavourable cropping practices. As such problems are today widespread in the tropics, there is an urgent need to characterize the biodiversity of tropical natural enemies, to identify what component of it is important for what crops, and develop means to protect and utilize it more effectively. Where we cannot predict in advance our future needs for natural enemies, as for the control of exotic pests, we must bear these needs in mind when deciding the fate of natural and agricultural habitats around the world.

References

Alves, S.B., Risco, S.H., Silveira Neto, S. and Machado Neto, R. (1984) Pathogenicity of nine isolates of *Metarhizium anisopliae* (Metsch.) Sorok. to *Diatraea saccharalis* (Fabr.). *Zeitschrift für angewandte Entomologie* 97, 403–406.

Askew, R.R. (1971) *Parasitic Insects*. Heinemann Educational Books, London.

Askew, R.R. and Shaw, M.R. (1986) Parasitoid communities: their size, structure and development. In: Waage, J.K. and Greathead, D.J. (eds), *Insect Parasitoids*. Academic Press, London, pp. 225–264.

Baretto, R.A. and Evans, H.C. (1988) Taxonomy of a fungus introduced into Hawaii for biological control of *Ageratina riparia* (Eupatorieae: Compositae), with observations on related weed pathogens. *Transactions of the British Mycological Society* 91, 81–97.

Bess, H.A., van den Bosch, R. and Haramoto, F.H. (1961) Fruit fly parasites and their activities in Hawaii. *Proceedings of the Hawaiian Entomological Society* 17, 367–378.

Burn, A.J. (1989) Long-term effects of pesticides on natural enemies of cereal crop pests. In: Jepson, P.C. (ed.), *Pesticides and Non-target Invertebrates*. Intercept, Andover, Hants, pp. 177–193.

Carl, K.P. (1989) Attributes of effective natural enemies, including identification of natural enemies for introduction purposes. In: King, E.G. and Jackson, R.D. (eds), *Proceedings of the Workshop on Biological Control of Heliothis, 11–15 November, 1985, New Delhi*. Far Eastern Regional Research Office, US Department of Agriculture, New Delhi, pp. 351–362.

DeBach, O., Rosen, D. and Kennett, C.E. (1971) Biological control of coccids by introduced natural enemies. In: Huffaker, C.B. (ed.), *Biological Control*. Plenum Press, New York, pp. 105–194.

Doyle, C.J., Hirst, M.L., Cory, J.S. and Entwistle, P.F. (1990) Risk assessment studies: detailed host range testing of wild-type cabbage moth, *Mamestra brassicae* (Lepidoptra: Noctuidae) nuclear polyhedrosis virus. *Applied and Environmental Microbiology* 56, 2704–2710.

van Emden, H.F. (1990) Plant diversity and natural enemy efficiency in agroecosystems. In: Mackauer, M., Ehler, L.E. and Roland, J. (eds), *Critical Issues in Biological Control*. Intercept, Andover, Hants, pp. 63–80.

Evans, H.C. (1982) Entomogenous fungi in tropical forest ecosystems: an appraisal. *Ecological Entomology* 7, 47–60.

Gauld, I.D. and Mound, L.A. (1982) Homoplasy and the delineation of holophyletic genera in some insect groups. *Systematic Entomology* 7, 73–86.

Godfray, H.C.J., Cock, M.J.W. and Holloway, J.D. (1987) An introduction to the Limacodidae and their bionomics. In: Cock, M.J.W., Godfray, H.C.J. and Holloway, J.D. (eds), *Slug and Nettle Caterpillars*. CAB International, Wallingford, pp. 1–8.

Greathead, D.J. and Waage, J.K. (1983) Opportunities for biological control of agricultural pests in developing countries. *World Bank Technical Paper* 11, i–xxx.

Hall, R.W., Ehler, L.E. and Bisarbi-Ershadi, B. (1980) Rate of success in classical biological control of arthropods. *Bulletin of the Entomological Society of America* 26, 111–114.

Hart, J.A. (1988) Rust fungi and host plant coevolution: do primitive hosts harbour primitive parasites? *Cladistics* 4, 339–366.

Hassell, M.P. and Waage, J.K. (1984) Host–parasitoid population dynamics. *Annual Review of Entomology* 29, 89–114.

Hawkins, B.A. (1988) Species diversity in the third and fourth trophic levels: patterns and mechanisms. *Journal of Animal Ecology* 57, 137–162.

Hawkins, B.A. (1990) Global patterns of parasitoid assemblage size. *Journal of Animal Ecology* 59, 57–72.

Hawksworth, D.L. (1991) The fungal dimension of biodiversity: magnitude, significance, and conservation. *Mycological Research* 95, 441–452.

Krieg, A., Huger, A.M., Langenbruch, G.A. and Schnetter, W. (1983) *Bacillus thuringiensis* var. *tenebrionis*: ein neuer, gegenueber Larven von Coleopteren wirksamer Pathotyp. *Zeitschrift für angewandte Entomologie* 96, 500–508.

Kuznezov, M. (1957) Number of species of ants in faunas of different latitudes. *Evolution* 11, 298–299.

Lawton, J.H. and Schroeder, D. (1978) Some observations on the structure of phytophagous insect communities: the implications for biological control. In: *Proceedings of the IVth International Symposium on Biological Control of Weeds. Gainesville, Florida.* University of Florida, Gainesville, pp. 57–73.

Letourneau, D.K. (1987) The enemies hypothesis: tritrophic interactions and vegetational diversity in tropical agroecosystems. *Ecology* 68, 1616–1622.

MacArthur, R.H. and Wilson, E.O. (1967) *The Theory of Island Biogeography*. Princeton University Press, Princeton.

McNaughton, S.J., Oesterheld, M., Frank, D.A. and Williams, K.J. (1989) Ecosystem-level patterns of primary productivity and herbivory in terrestrial habitats. *Nature* 341, 142–144.

Miller, J.C. (1980) Niche relationships among parasitic insects occurring in a temporary habitat. *Ecology* 61, 270–275.

Moscardi, F. (1989) Use of viruses for pest control in Brazil: The case of the nuclear polyhedrosis virus of the soyabean caterpillar, *Anticarsia gemmatalis*. *Memorias do Instituto Oswaldo Cruz* 84, 51–56.

Noyes, J.S. (1989) The diversity of Hymenoptera in the tropics with special reference to Parasitica in Sulawesi. *Ecological Entomology* 14, 197–207.

Perera, P.A.C.R., Hassell, M.P. and Godfray, H.C.J. (1988) Population dynamics of the coconut caterpillar (*Opisina arenosella* Walker) in Sri Lanka. *Bulletin of Entomological Research* 78, 479–492.

Price, P.W. (1980) *The Evolutionary Biology of Parasites*. Princeton University Press, Princeton.

Prior, C. (1989) Biological pesticides for low external–input agriculture. *Biocontrol News and Information* 10, 17–22.

Prior, C. and Greathead, D.J. (1989) Biological control of locusts: the potential for the exploitation of pathogens. *FAO Plant Protection Bulletin* 37, 37–48.

Samson, R.A., Evans, H.C. and Latge, J.P. (1988) *Atlas of Entomopathogenic Fungi*. Springer Verlag, Berlin.

Strong, D.R. (1974) Rapid asymptotic species accumulation in phytophagous insect communities: the pests of cacao. *Science* 185, 1064–1066.

Strong, D.R. and Levin, D.A. (1975) Species richness of the parasitic fungi of British trees. *Proceedings of the National Academy of Sciences, USA* 72, 2116–2119.

Strong, D.R., Lawton, J.H. and Southwood, T.R.E. (1984) *Insects on Plants: community patterns and mechanisms.* Blackwell Scientific Publications, Oxford.

Tisdell, C.A. (1990) Economic impact of biological control of weeds and insects. In: Mackauer, M., Ehler, L.G. and Roland, J. (eds), *Critical Issues in Biological Control.* Intercept, Andover, Hants, pp. 301–316.

Voegele, J.M., Agounke, D. and Moore, D. (1991) Biological control of fruit tree mealybug *Rastrococcus invadens* in Togo: a preliminary sociological and economic evaluation. *Tropical Pest Management*, in press.

Waage, J.K. and Greathead, D.J. (1988) Biological control: challenges and opportunities. *Philosophical Transactions of the Royal Society of London, B*, 318, 111–128.

World Health Organization (1982) *Biological Control of Vectors of Disease. Sixth Report of the WHO Expert Committee on Vector Biology and Control.* [Technical Report Series No. 679.] World Health Organization, Geneva.

Discussion*

Leslie: Herbicidal change or destruction by modern land-use practices induces species diversity. It also exposes previously unrecognized diversity. For example, unimodal crop protection has exposed intraspecies diversity in pests and pathogens whose recognition then becomes the basis for the construction of integrated pest management systems. This suggests that positive retention and use of biodiversity requires an iterative process of provoking the recognition of diversity through unimodal-type influences, coupled with a purposeful search for useful diversity in natural habitats.

Perfect: This is difficult to answer, and I have been wrestling with a similar issue. I think what we are doing by simplifying a system in physical terms is pushing more diversity in functional terms within a particular morphotype. The variation within a morphospecies in a diverse situation, especially at the genetic level, appears to be less than for the same morphospecies in simplified systems, such as a unimodal habitat. We are forcing out the variation in response to the particular crops and focusing on the variability within organisms, with consequent refining of taxonomic concepts. The biotypes of brown planthopper are a good example, the rice genotype determining the biotypes developed (Claridge, Chapter 15); an iterative relationship exists.

*Includes discussion related to Chapter 12

Holloway: I wonder if the concentration of diversity in components of a simplified systems is a fractal effect, a difference of scale. It would be of interest to know if there are comparable patterns of diversity in terms of genome distribution in a diverse natural system on the one hand, and within the variability of a crop or pest on the other.

Fourteen

Increasing Biodiversity to Improve Insect Pest Management in Agro-ecosystems

M.A. Altieri, *Division of Biological Control, University of California, Berkeley, California 94720, USA*

ABSTRACT: Agriculture implies the simplification of biodiversity, reaching an extreme form in crop monocultures, and the production of an artificial ecosystem requiring constant human intervention. The worsening of most pest problems is linked to the expansion of crop monoculture at the expense of natural vegetation. The ecological basis for the maintenance of biodiversity in agriculture, and the role it can play in restoring the ecological balance of agro-ecosystems, is analysed. Biodiversity performs a variety of renewal processes and ecological services in agro-ecosystems; when these are lost the costs can be significant. The ways in which biodiversity can contribute to the design of pest-stable agro-ecosystems are discussed. Natural enemy abundance can be enhanced by providing alternative host prey, food, refuges, and maintaining acceptable populations. The available experimental evidence is reviewed, and a variety of cases is described. While insect communities in agro-ecosystems can be stabilized by constructing vegetational architecture which supports natural enemies, each situation must be assessed separately, and long-term management strategies need to be developed with regard also to socio-economic factors.

Introduction

Biodiversity refers to all species of plants, animals, and microorganisms existing and interacting within an ecosystem (McNeely *et al.*, 1990). Global threats to biodiversity should not be foreign to agriculturalists, as agriculture

The Biodiversity of Microorganisms and Invertebrates: Its Role in Sustainable Agriculture.
Edited by D.L. Hawksworth. © CAB International 1991.

which covers about 25–30% of the world land area, is perhaps one of the main activities affecting biological diversity.

One effect results from the fact that agriculture implies the simplification of the structure of the environment over vast areas, replacing nature's diversity with a small number of cultivated plants and domesticated animals. In fact the world's agricultural landscapes are planted to only some 12 species of grain crops, 23 vegetable crop species and about 35 fruit and, nut crop species (Fowler and Mooney, 1990). That is no more than 70 plant species spread over approximately 1 440 million hectares of presently cultivated land in the world (Brown and Young, 1990). A sharp contrast with the diversity of plant species found within 1 hectare of a tropical rain forest which typically contains over 100 tree species (Myers, 1984).

The process of biodiversity simplification reaches an extreme form in agricultural monocultures. Indeed, modern agriculture is shockingly dependent on a handful of varieties for its major crops. For example, in the USA, 60–70% of the total acreage is planted to two to three bean varieties, 72% to four potato varieties and 53% to three cotton varieties (National Academy of Sciences, 1972). Researchers have repeatedly warned about the extreme vulnerability associated with this genetic uniformity.

The net result of biodiversity simplification for agricultural purposes is an artificial ecosystem that requires constant human intervention. Commercial seed-bed preparation and mechanized planting replace natural methods of seed dispersal; chemical pesticides replace natural controls on populations of weeds, insects and pathogens; and genetic manipulation replaces natural processes of plant evolution and selection. Even decomposition is altered as plant growth is harvested and soil fertility maintained, not through nutrient recycling, but with fertilizers.

Another way in which agriculture affects biodiversity is through the externalities associated with the intensive agrochemical and mechanical technology used to boost crop production. In the USA about 17.8 million tons of fertilizers are used in grain production systems, and about a half billion pounds of pesticides are annually applied to farmlands. Although these inputs have boosted crop yields, their undesirable environmental effects are undermining the sustainability of agriculture. Environmental and social costs associated with pesticide use are crudely estimated to reach about US$850 million in the USA (Pimentel *et al.*, 1980). About 27% of the irrigated land in the USA is damaged by salinization due to excessive or improper irrigation, and because of lack of rotations and insufficient vegetation cover, soil erosion levels average $18\,t\,ha^{-1}\,year^{-1}$ in USA croplands, well above the acceptable threshold (Brown and Young, 1990).

Nowhere are the consequences of biodiversity reduction more evident than in agricultural pest management. The instability of agro-ecosystems becomes manifest as the worsening of most insect pest problems is increasingly linked to the expansion of crop monocultures at the expense of the natural vegetation, thereby decreasing local habitat diversity (Altieri and Letourneau, 1982; Flint and Roberts, 1988). Plant communities that are modified to meet the special needs of humans become subject to heavy pest damage, and generally the more intensely such communities are modified, the more abundant and serious the pests. The inherent self-regulation characteristics of natural communities are lost when humans modify such communities through the shattering of the fragile web of community interactions (Turnbull, 1969). This breakdown can be repaired by restoring the shattered elements of community homeostasis, through the addition or enhancement of biodiversity.

In this contribution I explore the relationships between biodiversity and pest population dynamics in modern cropping systems subjected to diversification schemes. Instead of relying on the many reviews that comprehensively cover the subject of plant diversity and pest management in agro-ecosystems (van den Bosch and Telford, 1964; Dempster and Coaker, 1974; Williams, 1974; Murdoch, 1975; Litsinger and Moody, 1976; Perrin, 1977; Perrin and Phillips, 1978; Perrin, 1980; Price *et al.*, 1980; Cromartie, 1981; Altieri and Letourneau, 1982, 1984; Kareiva, 1983; Risch *et al.*, 1983; Altieri, 1984; Altieri and Liebman, 1986), instead I analysed the ecological basis for the maintenance of biodiversity in agriculture, and the role it can play in restoring ecological balance in agro-ecosystems. My theoretical assertions are illustrated with case studies of agricultural systems where biodiversity restoration and/or enhancement has resulted in effective pest regulation.

The ecological role of biodiversity in agriculture

One of the most important reasons for maintaining natural biodiversity is that it provides the foundation for all agricultural plants and animals. The entire range of our domestic crops is derived from wild species that have been modified through domestication, selective breeding and hybridization. Most remaining world centres of diversity contain populations of variable and adaptable landraces as well as wild and weedy relatives of crops (Harlan, 1975). Many traditionally managed farming systems in the Third World constitute *in situ* repositories of native crop diversity (Altieri and Merrick, 1987). There is great concern today about crop genetic erosion in areas where small farmers are pushed by agricultural modernization to adopt varieties at the expense of traditional ones.

In addition to producing valuable plants and animals, biodiversity performs many ecological services. In natural ecosystems the vegetative cover of a forest or grassland prevents soil erosion, replenishes ground water and controls flooding by enhancing infiltration and reducing water runoff. In agricultural systems biodiversity performs ecosystem services beyond the production of food, fibre, fuel, and income. Examples include the recycling of nutrients, control of local microclimate, regulation of local hydrological

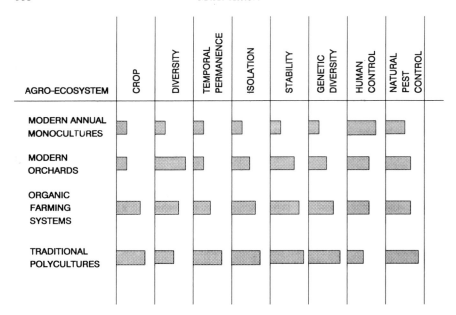

Figure 14.1. Ecological patterns of contrasting agro-ecosystems.

processes, regulation of the abundance of undesirable organisms, detoxification of noxious chemicals, etc. These renewal processes and ecosystem services are largely biological, therefore their persistence depends upon the maintenance of biological diversity. When these natural services are lost due to biological simplification, the economic and environmental costs can be significant. Economically in agriculture, these burdens include the need to supply crops with costly external inputs, as agro-ecosystems deprived of basic regulating functional components lack the capacity to sponsor their own soil fertility and pest regulation. Often the costs involve a reduction in the quality of life due to decreased soil, water and food quality when pesticide and/or nitrate contamination occurs.

Biodiversity and the design of pest-stable agro-ecosystems

Across the world, agro-ecosystems differ in age, diversity, structure and management. In fact there is great variability in basic ecological and agronomic patterns among the various dominant agro-ecosystems (Fig. 14.1). In general, the degree of biodiversity in agro-ecosystems depends on four main characteristics (Southwood and Way, 1970):

1. The diversity of vegetation within and around the agro-ecosystem.
2. The permanence of the various crops within the agro-ecosystem.

3. The intensity of management.
4. The extent of the isolation of the agro-ecosystem from natural vegetation.

In general, agro-ecosystems that are more diverse, more permanent, isolated and managed with low input technology (i.e. agroforestry systems, traditional polycultures) take fuller advantage of work usually done by ecological processes associated with higher biodiversity than highly simplified, input-driven and disturbed systems (i.e. modern vegetable monocultures and orchards).

All agro-ecosystems are dynamic and subject to different levels of management so that the crop arrangements in time and space are continually changing in the face of biological, cultural, socio-economic, and environmental factors. Such landscape variations determine the degree of spatial and temporal heterogeneity characteristic of agricultural regions, which in turn may or may not benefit the pest protection of particular agro-ecosystems. Thus, one of the main challenges facing agro-ecologists today is identifying the types of heterogeneity (either at the field or regional level) that will yield desirable agricultural results (i.e. pest regulation), given the unique environment and entomofauna of each area. This challenge can only be met by further analysing the relationship between vegetation diversification and the population dynamics of herbivore species, in the light of the diversity and complexity of agricultural systems. A hypothetical pattern in pest regulation according to agro-ecosystem temporal and spatial diversity is depicted in Fig. 14.2. Whether this pattern holds across regions and systems is unclear; however, our current understanding has been significantly advanced during the last two decades by numerous studies elucidating the ecological mechanisms that explain the ways in which plant diversity in agro-ecosystems influences the stability and diversity of the herbivore community (Altieri and Letourneau, 1982, 1984).

Although herbivores vary widely in their response to crop distribution, abundance, and dispersion, the majority of agro-ecological studies show that structural (i.e. spatial and temporal crop arrangement) and management (i.e. crop diversity, input levels, etc.) attributes of agro-ecosystems influence herbivore dynamics (Fig. 14.3). Several of these attributes are related to biodiversity and most amenable to management (i.e. crop sequences and associations, weed diversity, genetic diversity, etc.). Most experiments that have mixed other plant species with the primary host of a specialized herbivore showed that in comparison with diverse crop communities, simple crop communities have greater population densities of specialist herbivores (Root, 1973; Bach, 1980*a*, *b*; Risch, 1981). In these systems, herbivores exhibit greater colonization rates, greater reproduction, less tenure time, less disruption of host finding and enhanced mortality by natural enemies.

The two hypotheses that have been proposed to explain lower herbivore abundance in polycultures, the resource concentration hypothesis, and the enemies hypothesis (Root, 1973), identify key mechanisms of pest

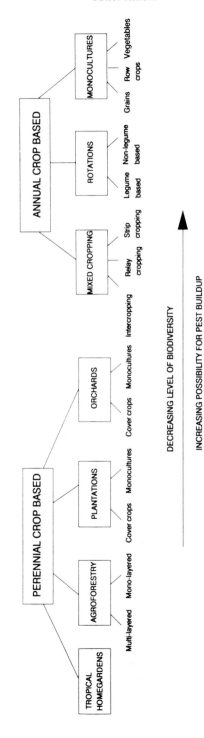

Figure 14.2. An hypothetical biodiversity gradient in a range of dominant agro-ecosystems.

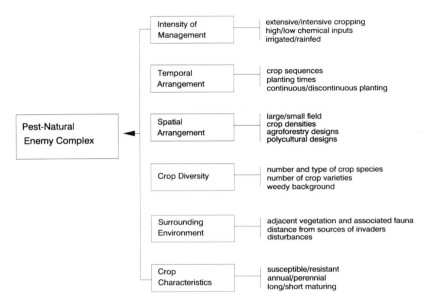

Figure 14.3. Spatial and temporal features of agro-ecosystems affecting the level of biodiversity and the dynamics of pest–natural enemy complexes.

regulations in polycultures, differences in mechanisms between cropping systems, and those plant assemblages which enhance regulatory effects and those which do not, and under what management and agro-ecological circumstances.

From a practical standpoint, it is easier to design insect manipulation strategies in polycultures using the elements of the natural enemies hypothesis than those of the resource concentration hypothesis; this is mainly because we cannot yet identify the ecological situations or life history traits that make some pests sensitive (i.e. their movement is affected by crop-patterning) and others insensitive to cropping patterns (Kareiva, 1986). Crop monocultures are difficult environments in which to induce the efficient operation of beneficial insects because these systems lack adequate resources for the effective performance of natural enemies, and because of the disturbing cultural practices often utilized in such systems. Polycultures already contain certain specific resources provided by plant diversity, and are usually not disturbed with pesticides (especially when managed by resource-poor farmers who cannot afford high-input technology). They are also more amenable to manipulation. Thus by replacing or adding diversity to existing systems, it may be possible to exert changes in habitat diversity that enhance natural enemy abundance and effectiveness by (van den Bosch and Telford, 1964; Altieri and Letourneau, 1982; Powell, 1986):

1. Providing alternative host prey at times of pest-host scarcity.
2. Providing food (pollen and nectar) for adult parasitoids and predators.
3. Providing refuges for overwintering, nesting, and so on.
4. Maintaining acceptable populations of the pest over extended periods to ensure continued survival of beneficial insects.

The specific resulting effect or the strategy to use will depend on the species of herbivores and associated natural enemies, as well as on properties of the vegetation, the physiological condition of the crop, or the nature of the direct effects of particular plant species (Letourneau, 1987). In addition, the success of enhancement measures can be influenced by the scale upon which they are implemented (i.e. field scale, farming unit, or region) since field size, within-field and surrounding vegetation composition, and level of field isolation (i.e. the distance from a source of colonizers) will all affect immigration rates, emigration rates, and the effective tenure time of a particular natural enemy in a crop field.

Perhaps one of the best strategies to increase the effectiveness of predators and parasitoids is the manipulation of non-target food resources (i.e. alternate host–prey and pollen–nectar) (Rabb *et al.*, 1976). Here it is not only important that the density of the non-target resource be high to influence enemy populations, but that the spatial distribution and temporal dispersion of the resource be also adequate. Proper manipulation of the non-target resource should result in the enemies colonizing the habitat earlier in the season than the pest, and frequently encountering an evenly distributed resource in the field, thus increasing the probability of the enemy to remain in the habitat and reproduce (Andow and Risch, 1985). Certain polycultural arrangements increase, and others reduce, the spatial heterogeneity of specific food resources; thus particular species of natural enemies may be more or less abundant in a specific polyculture. These effects and responses can only be determined experimentally across a whole range of agro-ecosystems. The task is indeed overwhelming, since enhancement techniques must necessarily be site-specific.

An appraisal of the experimental evidence

The literature is full of examples of experiments documenting that the diversification of cropping systems often leads to reduced herbivore populations. The published studies suggest that the more diverse the agro-ecosystem and the longer this diversity remains undisturbed, the more internal links develop to promote greater insect stability. It is clear, however, that the stability of the insect community depends not only on its trophic diversity, but on the actual density-dependence nature of the trophic levels (Southwood and Way, 1970). In other words, stability will depend on the precision of the response of any particular trophic link to an increase in the population from a lower level.

Although most experiments have documented population trends in single versus complex habitats, a few have concentrated on elucidating the nature and dynamics of the trophic relationships between plant–herbivore and natural enemies in diversified agro-ecosystems. In an analysis of 150 documented comparisons of herbivore abundance in monocultures and crop mixtures, Risch *et al.* (1983) discovered that 53% of the 148 herbivore species examined were less abundant in the diverse system than in monocultures. Several lines of studies have developed, which are considered in turn below.

Crop–weed–insect interaction studies

The evidence indicates that weeds influence the diversity and abundance of insect herbivores and associated natural enemies in crop systems. Certain weeds (mostly Umbelliferae, Leguminosae and Compositae) play an important ecological role by harbouring and supporting a complex of beneficial arthropods that assist in suppressing pest populations (Altieri *et al.*, 1977; Altieri and Whitcomb, 1979, 1980).

Insect dynamics in annual polycultures

Overwhelming evidence suggests that polycultures support a lower herbivore load than monocultures. One factor explaining this trend is that relatively more stable natural enemy populations can persist in polycultures due to the more continuous availability of food sources and microhabitats (Letourneau and Altieri, 1983; Helenius, 1989). The other possibility is that specialized herbivores are more likely to find and remain on pure crop stands, which provide concentrated resources and monotonous physical conditions (Tahvanainen and Root, 1972).

Herbivores in complex perennial crop systems

Most of these studies have explored the effects of the manipulation of ground cover vegetation on insect pests and associated enemies. The data indicates that orchards with a rich floral undergrowth exhibit a lower incidence of insect pests than clean cultivated orchards, mainly because of an increased abundance and efficiency of predators and parasitoids (Altieri and Schmidt, 1985). In some cases ground cover directly affects herbivore species which discriminate among trees with and without cover beneath.

The effects of adjacent vegetation

These studies have documented the dynamics of colonizing insect pests that invade field-crops from edge vegetation, especially when the vegetation is botanically related to the crop. A number of studies document the importance of adjoining wild vegetation in providing alternate food and habitat to natural enemies which move into nearby crops (van Emden, 1965; Wainhouse and Coaker, 1981; Altieri and Schmidt, 1986a).

Some relatively old case studies

Despite all the above experimental studies, we still have not been able to develop a predictive theory that enables us to determine what elements of biodiversity should be retained, added, or eliminated, to enhance natural pest control. In a few cases the simple addition of one element of diversity is all that is needed to ensure the biological control of a pest species (i.e. incorporating blackberries in vineyards to control *Erythroneura* leafhoppers in California; Doutt and Nakata, 1973). At times, all that is needed is to halt insecticide treatments to restore ecosystem regulating functions, as was the case in Costa Rican banana plantations; after 2 years of non-treatment, major insect pests decreased and many former insect pests nearly disappeared, and after 10 years most pests were under total biological control (Stephens, 1984). Recent studies comparing the arthropod fauna in organic and conventional farming systems, confirm the benefits of pesticide removal on the diversity of beneficial arthropods on the foliage and soils (Paoletti *et al.*, 1990).

We also know little to suggest whether at a more regional level (i.e. at the level of agro-ecosystem mosaics interspersed among natural vegetation) pest problems will diminish due to the spatial and temporal heterogeneity of agricultural landscapes. Some studies suggest that the vegetational settings associated with particular field crops influence the kind, abundance, and time of arrival of herbivores and their natural enemies (Gut *et al.*, 1982; Altieri and Schmidt, 1986b).

In perennial orchards (i.e. pear and apple) at temperate latitudes, a diverse complex of predators usually develops early in the season in orchards surrounded by woodlands. In these systems, the main pests (e.g. *Psylla pyricola, Cydia pomonella*) are quickly reduced and maintained at low levels throughout the season. In contrast, early season predators are absent in more extensive commercial orchards, and therefore pest pressure is more intense (Croft and Hoyt, 1983). Assuming that these trends also occur in the tropics, it would be expected that in certain tropical agro-ecosystems (i.e. shifting cultivation in the lowland tropics), forest and bush fallows have potential value in controlling pests. Clearing small plots in a matrix of secondary forest vegetation may permit the easy migration of natural enemies from the surrounding jungle (Matteson *et al.*, 1984). This positive potential role of natural vegetation on biological control can, however, be expected to change in view of current deforestation rates and modernization trends toward commercial monocultures.

One of the unique cases exploring the relationship between landscape ecology and insect pests comes from a 10-year experiment conducted near Waco, Texas, in 1929–49, by the Bureau of Entomology and Plant Quarantine in cooperation with the Soil Conservation Service. Entomologists measured the effects of new farming and soil conservation methods on populations of beneficial and pest insects in cotton (de Loach, 1970). About 600 acres of upland farmland was divided into adjacent areas of 300 acres each, designated Y and W (Fig. 14.4). In both, the old farming practices were continued for 4 years, through 1942, while pretreatment counts were made. New conservation methods then were begun in Y, while the old practices were continued in W. By the old practice, cotton occupied the largest acreage, followed closely by corn. There was also a substantial acreage of oats and pasture and a little sorghum; these crops occupied nearly 100% of the land area (Fig. 14.4). By the new practices, several acres of clover were planted alone or overseeded in the oats, some grass areas were added, and the land was terraced. Insecticides were not used in any fields in the two areas during the experiment, so that the effect of cultural methods alone could be measured. The new conservative practices resulted in a reduction in numbers of cotton pest insects and a reduction in the percentage of damaged squares and bolls. No attempt was made at elucidating the mechanisms explaining such reductions, although it is assumed that natural enemies were highly favoured by the new landscape designs, thus resulting in enhanced pest mortality. This study also gives credence to emerging approaches that suggest the importance of the landscape as a level of organization of processes such as dispersion of plants, arthropod movement and nutrient flow (Paoletti *et al.*, 1990). Since agriculture is a major force shaping landscape structure and dynamics, it is useful to examine the relationships between arthropods and vegetation patterns at the landscape ecological level.

In another large-scale situation, two 10-hectare experimental fields were compared in south-eastern Missouri (de Loach, 1970). One half-field was solidly planted to cotton, and the other half was strip-planted, with seven strips of cotton 16 rows wide, each with a strip of alfalfa on one side and either corn or oats and then soyabeans on the other (Fig. 14.5). Three predator species, *Orius insidiosus*, *Hippodamia convergens*, and *Coleomegilla maculata*, were highly abundant in the diverse plot. Populations of these predators probably developed in the strips of oats, alfalfa, and corn, and then moved into the strips of cotton. Densities of bollworm larvae were generally low. Other phytophagous insects were abundant at times, especially thrips, bugs,

M.A. Altieri

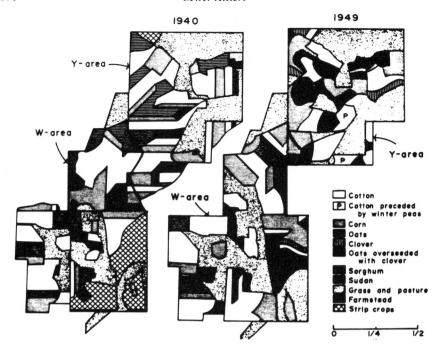

Figure 14.4. Agricultural landscape designs from 1940–49 in Waco, Texas, in areas without soil conservation practices (W) and with soil conservation and diversification practices (Y) (adapted from de Loach, 1970).

leafhoppers, and whiteflies, but they never reached economically damaging levels.

Placing eggs of the cabbage looper (*Trichoplusia ni*) in the plots did not always yield significant differences in predation pressure, but when significant differences between plots were found, predation was always higher in the more diversified habitat. Although not consistent throughout the two plots, in the west side of the fields damaged bolls in the strip-planted plots were 50% lower than in the solid planted plot.

Later experiments have corroborated the above results and have better defined some of the trophic interactions involved. Fye (1971) described a strip-crop system in which predators increased on aphids and greenbugs in grain sorghum and then migrated into adjacent cotton as the sorghum matured early in relation to the peak cotton fruiting period. The interplanting of sorghum and cotton greatly facilitated the interchange of *Collops* beetles, Coccinellidae, Chrysopidae, and *Geocoris* species. The density of predators was increased in cotton fields for a considerable distance from stripped cropped sorghum (Burleigh *et al.*, 1973).

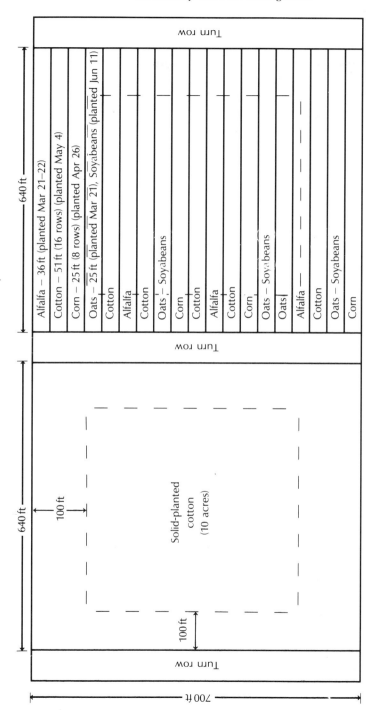

Figure 14.5. Experimental layout and arrangement of solid-planted cotton and strip-planted cotton plots in Missouri (adapted from de Loach, 1970).

Conclusions

Biodiversity in agro-ecosystems can be as varied as the various crops, weeds, arthropods, microorganisms, etc., involved, according to geographical location, climatic, edaphic, and human factors. Complementary interactions between the various biotic components can also be of a multiple nature. Some of these interactions can have positive or negative, direct or indirect, effects on the biological control of specific crop pests. The exploitation of these interactions in real situations involves agro-ecosystem design and management, and requires an understanding of the numerous relationships between plants, herbivores and natural enemies.

The experimental evidence suggests that biodiversity can be used for improved pest management. There is no doubt that it is possible to stabilize the insect communities of agro-ecosystems, by designing and constructing vegetational architectures which support populations of natural enemies or have direct deterrent effects on pest herbivores. What is difficult is that each agricultural situation must be assessed separately, since herbivore–enemy interactions will vary significantly depending on insect species, location and size of the field, plant composition, the surrounding vegetation, and cultural management. One can only hope to elucidate the ecological principles governing herbivore dynamics in complex systems, but the biodiversity designs necessary to achieve herbivore regulation will depend on the agro-ecological conditions and socio-economic restrictions of each area. In this regard, farmers' needs and preferences must be fully considered if the adoption of new designs is expected. New designs will be most attractive if, in addition to pest regulation, diversification schemes offer benefits in terms of overyielding, increased soil fertility, decreased weed competition and diseases, and an evening out of labour demands.

Long-term maintenance of diversity requires a management strategy that considers regional biogeography and landscape patterns above local concerns of production. So far, profit motivations rather than environmental considerations have influenced the forces shaping the structure of agriculture, and it is the very nature of such agricultural structure and associated policies that lead to environmental degradation and biodiversity simplification in agriculture (Buttel, 1980). For example, biased policies that encourage farm size expansion favour specialized production which in turn leads to monoculture, a production system that relies on chemical inputs and abandonment of biodiversity conservation practices such as rotations, maintenance of hedgerows, or multiple cropping. This is why I have repeatedly questioned whether the pest problems of modern agriculture can be ecologically alleviated within the context of the present capital-intensive structure of agriculture (Altieri, 1987). It is clear that many environmental problems of modern agriculture are rooted within the economic/policy structure. Therefore, it is crucial to consider major social change, research and extension reorientation, land reform, and redesign of machines, in the agricultural sector to increase the possibilities of improved pest control through the management of biodiversity.

References

Altieri, M.A. (1984) Diversification of agricultural landscapes: a vital element to pest control in sustainable agriculture. In: Edens, T. (ed.), *Proceedings of the Conference on Sustainable Agriculture and Integrated Farming Systems.* Michigan State University Press, East Lansing, pp. 166–184.

Altieri, M.A. (1987) *Agroecology: the scientific basis of alternative agriculture.* Westview Press, Boulder.

Altieri, M.A. and Letourneau, D.K. (1982) Vegetation management and biological control in agroecosystems. *Crop Protection* 1, 405–430.

Altieri, M.A. and Letourneau, D.K. (1984) Vegetation diversity and pest outbreaks. *Critical Reviews in Plant Sciences* 2, 131–169.

Altieri, M.A. and Liebman, M. (1986) Insect, weed and plant disease management in multiple cropping systems. In: Francis, C.A. (ed.), *Multiple Cropping Systems.* MacMillan, New York, pp. 183–218.

Altieri, M.A. and Merrick, L.C. (1987) *In situ* conservation of crop genetic resources through maintenance of traditional farming systems. *Economic Botany* 4, 86–96.

Altieri, M.A. and Schmidt, L.L. (1985) Cover crop manipulation in northern California apple orchards and vineyards: effects on arthropod communities. *Biological Agriculture and Horticulture* 3, 1–24.

Altieri, M.A. and Schmidt, L.L. (1986a) Cover crops affect insect and spider populations in apple orchards. *California Agriculture* 40, 15–17.

Altieri, M.A. and Schmidt, L.L. (1986b) The dynamics of colonizing arthropod communities at the interface of abandoned, organic and commercial apple orchards and adjacent woodland habitats. *Agriculture Ecosystems and Environment* 16, 29–43.

Altieri, M.A. and Whitcomb, W.H. (1979) The potential use of weeds in the manipulation of beneficial insects. *Horticultural Science* 14(1), 12–18.

Altieri, M.A. and Whitcomb, W.H. (1980) Weed manipulation for insect management in corn. *Environmental Management* 4, 483–489.

Altieri, M.A., Schoonhoven, A.v. and Doll, J.D. (1977) The ecological role of weeds in insect pest management systems: a review illustrated with bean (*Phaseolus vulgaris* L.) cropping systems. *PANS* 23, 195–205.

Andow, D. and Risch, J.J. (1985) Predation in diversified agroecosystems: relations between a coccinellid predator *Coleomegilla maculata* and its food. *Journal of Applied Ecology* 22, 357–372.

Bach, C.E. (1980a) Effects of plant density and diversity on the population dynamics of a specialist herbivore, the striped cucumber beetle, *Acalymma vittata* (Fab). *Ecology* 61, 1515–1530.

Bach, C.E. (1980b) Effects of plant diversity and time of colonization on an herbivore–plant interaction. *Oecologia* 44, 319–326.

Brown, L.R. and Young, J.E. (1990) Feeding the world in the nineties. In: Brown, L.R. (ed.), *State of the World*. W.W. Norton, New York, pp. 106–124.

Burleigh, J.G., Young, J.H. and Morrison, R.D. (1973) Strip cropping's effect on beneficial insects and spiders associated with cotton in Oklahoma. *Environmental Entomology* 2, 281.

Buttel, F.H. (1980) Agricultural structure and rural ecology: toward a political economy of rural development. *Sociologia Ruralis* 20, 44–62.

Croft, B.A. and Hoyt, S.C. (1983) *Integrated Management of Insect Pests of Pome and Stone Fruits*. John Wiley, New York.

Cromartie, W.J. (1981) The environmental control of insects using crop diversity. In: Pimentel, D. (ed.), *CRC Handbook of Pest Management*. CRC Press, Boca Raton, pp. 223–251.

de Loach, C.J. (1970) The effect of habitat diversity on predation. In: Komarek, E.V. (ed.), *Proceedings of the Tall Timbers Conference on Ecological Animal Control by Habitat Management*. Tall Timbers Research Station, Tallahasse, 2, pp. 223–241.

Dempster, J.P. and Coaker, T.H. (1974) Diversification of crop ecosystems as a means of controlling pests. In: Jones, D.P. and Solomon, M.E. (eds), *Biology in Pest and Disease Control*. John Wiley, New York, pp. 106–114.

Doutt, R.L. and Nakata, J. (1973) The *Rubus* leafhopper and its egg parasitoid: an endemic biotic system useful in grape pest management. *Environmental Entomology* 2, 381–386.

Flint, M.L. and Roberts, P.A. (1988) Using crop diversity to manage pest problems: some California examples. *American Journal of Alternative Agriculture* 3, 164–167.

Fowler, C. and Mooney, P. (1990) *Shattering. Food, Politics and the Loss of Gene Diversity*. University of Arizona Press, Tucson.

Fye, R.E. (1971) Grain sorghum – a source of insect predators for insects on cotton. *Progress in the Agriculture of Arizona* 23, 12.

Gut, L.J., Jochums, C.E., Westigard, P.H. and Liss, W.J. (1982) Variation in pear psylla (*Psylla pyricola* Foerster) densities in southern Oregon orchards and its implications. *Acta Horitculturae* 124, 101–111.

Harlan, J.R. (1975) Our vanishing genetic resources. *Science* 188, 618–622.

Helenius, J. (1989) Intercropping, insect populations and pest damage: case study and conceptual model. *Report of the Department of Agricultural and Forestry Zoology, University of Helsinki* 14, 49.

Kareiva, P. (1983) Influence of vegetation structure on herbivore populations: resource concentration and herbivore movement. In: Denno, R. (ed.), *Variable Plants and Herbivores in Natural and Managed Systems*. Academic Press, New York, pp. 259–286.

Kareiva, P. (1986) Trivial movement and foraging by crop colonizers. In: Kogan, M. (ed.), *Ecological Theory and Integrated Pest Management Practice*. J. Wiley & Sons, New York, pp. 59–82.

Letourneau, D.K. (1987) The enemies hypothesis: tritrophic interactions and vegetational diversity in tropical ecosystems. *Ecology* 68, 1616–1622.

Letourneau, D.K. and Altieri, M.A. (1983) Abundance patterns of a predator, *Orius tristicolor* (Hemiptera: Anthocoridae), and its prey, *Frankliniella occidentalis* (Thysanoptera: Thripidae): habitat attraction in polycultures versus monocultures. *Environmental Entomology* 12, 1464–1469.

Litsinger, J.A. and Moody, K. (1976) Integrated pest management in multiple crop-ping systems. In: Sanchez, P.A. (ed.), *Multiple Cropping*. American Society of Agronomy, Madison, pp. 293–316.

McNeely, J.A., Miller, K.R., Reid, W.V., Mittermeier, R.A. and Werner, T.B. (1990) *Conserving the World's Biological Diversity*. International Union for Conservation of Nature and Natural Resources, Gland.

Matteson, P.C., Altieri, M.A. and Gagne, W.C. (1984) Modification of small farmer practices for better management. *Annual Review of Entomology* 29, 383–402.

Murdoch, W.W. (1975) Diversity, complexity, stability and pest control. *Journal of Applied Ecology* 12, 795–807.

Myers, N. (1984) *The Primary Source: Tropical Forests and our Future*. W.W. Norton, New York.

National Academy of Sciences (1972) *Genetic Vulnerability of Major Crops*. National Academy Press, Washington, DC.

Paoletti, M.G., Stinner, B.R. and Lorenzoni, G.T. (eds) (1990) *Agricultural Ecology and Environment*. Elsevier, Amsterdam.

Perrin, R.M. (1977) Pest management in multiple cropping systems. *Agro-Ecosystems* 3, 93–118.

Perrin, R.M. (1980) The role of environmental diversity in crop protection. *Protection Ecology* 2, 77–114.

Perrin, R.M. and Phillips, M.L. (1978) Some effects of mixed cropping on the population dynamics of insect pests. *Entomologia Experimentalis et Applicata* 24, 385–393.

Pimentel, D., Andow, D., Dyson-Hudson, R., Gallahan, D., Jacobson, S., Irish, M., Kroop, S., Moss, A., Schreiner, I., Shepard, M., Thompson, T. and Vinzant, B. (1980) Environmental and social costs of pesticides: a preliminary assessment. *Oikos* 34, 126–140.

Powell, W. (1986) Enhancing parasitoid activity in crops. In: Waage, J.W. and Greathead, D.J. (eds), *Insect Parasitoids*. Academic Press, London.

Price, P.W., Bouton, C.E., Gross, P., McPherson, J.A., Thompson, B.N. and Weise, A.E. (1980) Interaction among three trophic levels: influence of plants on inter-action between insect herbivores and natural enemies. *Annual Review of Ecology and Systematics* 11, 41–65.

Rabb, R.L., Stinner, R.E. and van den Bosch, R. (1976) Conservation and augmen-tation of natural enemies. In: Huffaker, C.B. and Messenger, P.S. (eds), *Theory and Practice of Biological Control*. Academic Press, New York, pp. 233–253.

Risch, S. (1981) Insect herbivore abundance in tropical monocultures and poly-cultures: an experimental test of two hypotheses. *Ecology* 62, 1325–1340.

Risch, S.J., Andow, D. and Altieri, M.A. (1983) Agroecosystem diversity and pest control: data, tentative conclusions and new research directions. *Environmental Entomology* 12, 625–629.

Root, R.B. (1973) Organization of a plant–arthropod association in simple and diverse habitats: the fauna of collards (*Brassicae oleraceae*). *Ecological Mono-graphs* 43, 95–124.

Southwood, T.R.E. and Way, M.J. (1970) Ecological background to pest manage-ment. In: Rabb, R.L. and Guthrie, F.E. (eds), *Concepts of Pest Management*. North Carolina State University, Raleigh, pp. 6–29.

Stephens, C.S. (1984) Ecological upset and recuperation of natural control of insect pests in some Costa Rican banana plantations. *Turrialba* 34, 101–105.

Tahvanainen, J.O. and Root, R.B. (1972) The influence of vegetational diversity on the population ecology of a specialized herbivore, *Phyllotreta cruciferae* (Coleoptera: Chrysomelidae). *Oecologia* 10, 321–346.

Turnbull, A.L. (1969) The ecological role of pest populations. *Proceedings of the Tall Timbers Conference on Ecological Animal Control by Habitat Management* 1, 219–232.

van den Bosch, R. and Telford, A.D. (1964) Environmental modification and biological control. In: DeBach, P. (ed.), *Biological Control of Insect Pests and Weeds*. Chapman and Hall, London, pp. 459–488.

van Emden, H.F. (1965) The role of uncultivated land in the biology of crop pests and beneficial insects. *Scientific Horticulture* 17, 121–126.

van Emden, H.F. (1990) Plant diversity and natural enemy efficiency in agroecosystems. In: MacKauer, M., Ehler, L. and Roland, J. (eds), *Critical Issues in Biological Control*. Intercept, Andover, pp. 63–80.

van Emden, H.F. and Williams, G.F. (1974) Insect stability and diversity in agroecosystems. *Annual Review of Entomology* 19, 455–475.

Wainhouse, D. and Coaker, T.H. (1981) The distribution of carrot fly (*Psila rosae*) in relation to the fauna of field boundaries. In: Thresh, J.M. (ed.), *Pests, Pathogens and Vegetation: The Role of Weeds and Wild Plants in the Ecology of Crop Pests and Diseases*. Pitman, Massachusetts, pp. 263–272.

Fifteen

Genetic and Biological Diversity of Insect Pests and their Natural Enemies

M.F. Claridge, *School of Pure and Applied Biology, University of Wales, P.O. Box 915, Cardiff CF1 3TL, UK*

ABSTRACT: The major unit of insect ecological diversity should be the biological species, most clearly defined by the quality of reproductive isolation. The techniques of biosystematics, or biotaxonomy, are multidisciplinary and designed to demonstrate reproductive isolation and genetic diversity in field populations.

The brown planthopper (*Nilaparvata lugens*), a major pest of tropical rice, is a well-studied example that demonstrates the difficulties of relying on simple concepts for an understanding of complex natural systems. Most insect pests of agriculture and forestry and their natural enemies need intensive biosystematic work if systems of integrated management for truly sustainable agriculture are to be achieved.

Top priority must be given to funding networks of appropriate institutions around the world and the targeting of major pests for research. Urgent allocation of international funding is essential.

Introduction

Ecologists have devised many indices by which to measure the biological diversity of ecosystems and communities. However, most now agree that for practical purposes the total number of species in any system, that is species richness, is both the simplest and the most useful index. Unfortunately, however, few ecologists have given much attention to the nature of the species category itself (Claridge, 1987). Clearly, any index of species diversity must be based on an agreed and if possible objective species concept.

The Biodiversity of Microorganisms and Invertebrates: Its Role in Sustainable Agriculture.
Edited by D.L. Hawksworth. © CAB International 1991.

Species concepts

The species concept employed for insects has usually been a purely morphological one for reasons of convenience because of the enormous numbers of species involved. However, the morphological species concept is clearly a subjective and arbitrary one. Thus, measures of diversity based on it must also be to some extent subjective and arbitrary.

Unlike most microorganisms, most insects are characterized by systems of biparental sexual reproduction. The associated breeding systems are thus such that high levels of variation are inevitably produced by genetic recombination. This contrasts with asexual and parthenogenetic organisms in which genetic variation is always less and sometimes very little. An important consequence for biparentally reproducing organisms is that the biological species, or biospecies, is the most useful and meaningful unit of diversity. The biological species, as developed by authors such as Dobzhansky (1937), Mayr (1942, 1963) and Cain (1954), depends essentially on the criterion of reproductive isolation. Such isolation is a genetic phenomenon and is not necessarily accompanied by clear morphological differentiation (Claridge, 1988). Thus, genetically isolated species may exist that show little obvious differentiation to the morphological taxonomist, but which may differ in important biological characteristics. Some of the best known examples of such sibling, or cryptic, species are found among the important groups of biting Diptera, including mosquitoes, black flies, sand flies, etc. Sibling species of these flies often differ vitally in such characteristics as host preference and ability to transmit particular disease organisms. Medical and veterinary entomologists therefore rarely rely on morphological characteristics for identifying critical species. Agricultural entomologists have been much slower to recognize such problems and because of this they undoubtedly regularly underestimate levels of genetic diversity.

The recognition of biological species is a vital first step in the understanding of insect pests, as indeed of other insects. The techniques required to recognize biological species are often referred to as biosystematics, or biotaxonomy, and are designed to test for reproductive isolation between populations in the field. Such tests rarely involve the direct observation of reproductive isolation, but more usually they rely on recognizing diagnostic markers correlated with the reproductive barrier. Traditionally such markers for insects are generally morphological ones. The morphological species will often correlate well with the biological species because the chosen characters are reliable indicators of reproductive isolation. Indeed, most basic taxonomic research relies on this. However, the phenotypic variation consequent upon sexual reproduction often causes great difficulties of interpretation.

Over the past 20–30 years, a variety of more definite genetic techniques have become available to aid in the identification of reproductive isolation. The oldest of these is indeed an extension of morphological taxonomy to incorporate morphological characters of the chromosomes themselves. Appropriate microscopical techniques allow a taxonomist access to a whole range of new characters and have often enabled definite identification of otherwise impossible reproductively isolated species. Classic examples are provided by the *Anopheles gambiae* complex of mosquito vectors of malaria in Africa. Mosquitoes, like so many Diptera, have chromosomes that are relatively easy to study in great detail. The enormous literature on the structure and banding patterns of chromosomes in species of *Drosophila* bears witness to the great value of such features. Unfortunately, in most other organisms, detailed studies of chromosome structure are not so easily possible.

Many newer techniques are now available for studying genotypic variation, including DNA sequencing, DNA hybridization, protein sequencing, and protein gel electrophoresis (Loxdale and den Hollander, 1989). These techniques are now widely used by evolutionary biologists to study genetic variation in natural populations, and provide powerful data to establish the existence or otherwise of gene flow between populations. Studies on major insect pests have demonstrated the existence of previously unsuspected sibling species (e.g. Menken and Ulenberg, 1987; Menken, 1989; Blackman *et al.*, 1989). Thus, techniques for the taxonomist are now much more varied than traditional gross morphological studies. Unfortunately, such new techniques are often not available for routine use in taxonomic laboratories.

Whatever type of marker has to be used in species taxonomy, whether it be morphological or biochemical, it is still only a marker of genetic, and therefore of reproductive, isolation. However, biparental organisms are characterized by behavioural mechanisms that help to ensure that mating takes place only between conspecific individuals and that therefore maintain reproductive isolation between species. They are what Mayr (1942) termed behavioural isolating mechanisms, and what Paterson (1985) prefers to term specific mate recognition systems. The behavioural interactions between males and females during mate finding and courtship involve the exchange of specific signals. They certainly function as recognition signals, and have the effect of maintaining reproductive isolation. Thus in practice the two concepts are very similar (Claridge, 1988).

Among insects, such mate recognition systems are extremely diverse, but may include olfactory, gustatory, visual, tactile, or acoustic signals, and usually probably various combinations of several. The study of such signal systems provides the best evidence for species status and actual and potential reproductive isolation between populations. Unfortunately, little attention has been given to such studies, but, where they have been made, fundamental differences in the interpretation of patterns of variation in the field have resulted (see below).

Intraspecific variation

The problem of variation between individuals and populations of biparental organisms is a major one for the taxonomist. As discussed above, a difficulty is to identify from the array of variability some characters that may be used for discriminating biological species. However, variation within species clearly contributes much of great importance to an understanding of biological diversity and a variety of categories are used by taxonomists and evolutionary biologists to describe it. Such terms as race, biotype, ecotype, subspecies, etc., are widely used, but rarely clearly defined (Claridge and den Hollander, 1983). Often there is a complete failure to differentiate between variation between spatially or geographically separate (allopatric) populations, and that between sympatric populations. Nowhere is this problem greater than among specialized feeders, such as insect herbivores and their parasitoids. The accurate determination of the status of populations of host-plant-associated insect herbivores is essential to an understanding of agricultural pest diversity. Often populations appear to differ only in their abilities to attack and colonize particular host-plants. Unfortunately, only very rarely are the appropriate resources available to solve such problems. They usually require, among others, the application of behavioural, biochemical, and multivariate morphometric methods.

Diversity in agro-ecosystems

The dramatic consequences of decreasing biological diversity in agro-ecosystems have been demonstrated unwittingly time and again by modern agricultural practices. The now well-known consequences of the indiscriminate use of broad-spectrum pesticides include the disruption of diverse communities and usually the dramatic reduction in both species diversity and total numbers of natural enemies present. The result has so often been the creation of new and intractable pests as a direct result of massive reductions in biological diversity.

Similarly, the tendency in modern agriculture has generally been to reduce the genetic diversity of crop plants by planting very large areas with a single favoured cultivar. Usually this has been to the exclusion of other cultivars and thus has had the effect of dramatically reducing genetic diversity in the system. The consequences are that, when pest outbreaks occur, they are often very severe and difficult to manage.

For these reasons a more ecological approach is advocated almost universally today for pest management. It is now commonplace to attempt to develop systems of integrated pest management based on ecologically sound principles, not only of population dynamics, but also of community structure and biological diversity. The major problem is that always these systems require first knowledge of the complex of organisms involved in each ecosystem. Particularly in tropical regions, the basic morphological taxonomy of important insect pests and their relatives has almost never been adequately researched so that detailed biosystematic work should be a top priority.

Pest diversity

An instructive example that illustrates many of the general problems discussed above is provided by *N. lugens*, the brown planthopper, a major pest of irrigated rice throughout eastern and southern Asia and parts of Australia.

Rice feeding populations

Nilaparvata lugens feeds on rice species and cultivars of *Oryza*. It became a major pest in tropical Asia only following the development of the improved high-yielding rice varieties, first at the International Rice Research Institute (IRRI), and their subsequent widespread use by farmers in the early 1970s (IRRI, 1979; Claridge, 1990a). The first varieties were soon very susceptible to *N. lugens* feeding damage, often resulting in massive areas of hopperburn and almost total yield loss.

The major strategy for control has been that of breeding resistant cultivars, work which was pioneered and still continues at IRRI (IRRI, 1979; Heinrichs *et al.*, 1985; Saxena and Khan, 1989). The widespread use in farmers' fields of varieties with known genes for resistance led to the evolution of virulent populations of the pest that were able to damage previously resistant varieties. Such populations, characterized by particular patterns of virulence, were termed 'biotypes' and categorized by a system of numbering.

The nature of these biotypes has provoked much controversy. The term biotype itself is confusing, because it has been used to refer to such contrasting biological phenomena as sibling species, geographical races, and individual morphs in a polymorphic population (Claridge and den Hollander, 1983; Diehl and Bush, 1984). Indeed, as Claridge and Morgan (1987) suggested, it might be wise to follow Diehl and Bush (1984) who suggest

> that future application of the term [biotype] be restricted to use as a temporary and provisional designation for cases where biological differences have been observed between organisms but where the genetic basis and evolutionary status of the differences have yet to be ascertained.

The only other well-known example of a biparental insect in which the biotype concept has been widely used is the Hessian-fly, *Mayetiola destructor*, a pest of wheat in North America (Everson and Gallun, 1980). This insect appears to conform to the gene-for-gene relationship. That is, a gene for virulence in the fly is associated with a gene for resistance in the wheat plant. Thus each biotype is characterized by a major gene which corresponds to the

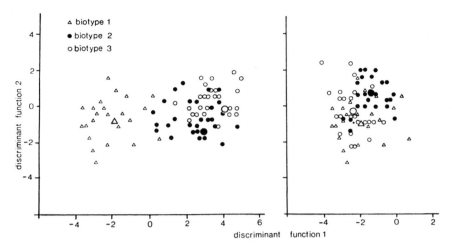

Figure 15.1. Plots of discriminant function 1 against discriminant function 2 for canonical analysis of morphometric characteristics of the IRRI biotypes of N. *lugens*. Biotype 1 on rice variety TNI, biotype 2 on Mudgo and biotype 3 on ASD7 (*left*), and biotypes 1, 2 and 3 all on TN1 after one generation (*right*) (after Claridge et al., 1984a).

ability to feed on wheat cultivars that incorporate the corresponding plant gene. In *N. lugens* there is no indication of such a gene-for-gene relationship, despite the claims by Saxena and Barrion (1987). There is indeed clear evidence for polygenic control of virulence (den Hollander and Pathak, 1981). Studies on individual insects of IRRI biotypes 1, 2 and 3, show that they are genetically very closely related populations that have adapted to rice varieties with particular genes for resistance (Claridge and den Hollander, 1982). Selection experiments have shown that one biotype may be converted to another under laboratory conditions after about 10 generations on an appropriate rice variety (Claridge and den Hollander, 1982).

It is thus clear that the IRRI biotypes differ biologically only in their characteristic patterns of virulence, with little evidence of other genetic divergence. However, Saxena and coworkers at IRRI have regularly claimed greater genetic differentiation of these biotypes. In particular, they have suggested that the biotypes may be identified by morphometric characters (Saxena and Rueda, 1982; Saxena and Barrion, 1985). We have indeed confirmed that it is possible to separate the IRRI biotypes when they are reared on their own normal laboratory cultivars, but when all were reared on the same susceptible cultivar (TNI) after only one generation no significant difference could be found between them (Fig. 15.1; Claridge et al., 1984). Thus, it is clear that the differences reported are predominantly environmentally induced and do not represent major genetic differentiation.

Most studies of *N. lugens* biotypes have been made with the IRRI cultures. These were originally established from field collections made in

Luzon and have been inbred for very many generations since the early 1970s. It is therefore dangerous to generalize from work on these to field populations. Nonetheless, the biotype numbering system has been widely extended to apply to field populations not only in the Philippines, but also to populations with similar gross patterns of virulence elsewhere in Asia. Indeed, Saxena and Rueda (1892) even stated that the morphometric characters used by them to identify the IRRI culture populations could be used to identify the 'same biotypes' in other regions. There are no published data to validate this most improbable suggestion.

We may conclude then that *N. lugens* is a labile species in terms of virulence characteristics and is capable of rapid adaptation to new host cultivars. This is well-illustrated by a study of six field populations collected from different rice varieties in Sri Lanka, each no more than 200 km apart (Claridge *et al.*, 1982). Each population differed significantly from each of the others in virulence characters and was best adapted to the variety from which it was collected.

Thus, though there certainly are some major differences in patterns of virulence in different parts of Asia (e.g. Seshu and Kaufman, 1980), a specific biotype terminology is likely to be misleading and cause confusion. Indeed, it superimposes a supposed simplicity on a complex and labile natural system. Apart from virulence characteristics, some genetic differentiation has been demonstrated between allopatric populations across the enormous natural range of distribution of *N. lugens* (Claridge *et al.*, 1985a). Hybridization experiments indicated some geographical differentiation in pre-mating reproductive barriers. All planthoppers and their relatives are characterized by the exchange of substrate-transmitted acoustic signals in mate location and courtship (Claridge, 1985). Analyses of these signals for *N. lugens* showed that differences in the rate of repetition of pulsed patterns in the calls were primarily responsible for observed levels of reproductive isolation between some populations (Claridge *et al.*, 1984b, 1985a). Indeed, analyses of acoustic signals provide the most subtle means yet discovered for determining genetic divergence between populations of *N. lugens*.

Weed associated populations

Populations indistinguishable in morphology from *N. lugens* are known to feed and breed in the field on the weed grass *Leersia hexandra* throughout tropical Asia and Australasia (Claridge *et al.*, 1985b; Claridge, 1990b). *L. hexandra* is related to *Oryza* and grows abundantly in drainage channels, ponds and reservoirs near paddy fields as well as in natural swamps and at the margins of lakes, often together with wild species of *Oryza*. In mate choice experiments insects showed strong, and often complete preference for mating with others from their own host. When transferred to the other host, insects died rapidly and did not survive to reproduce. However,

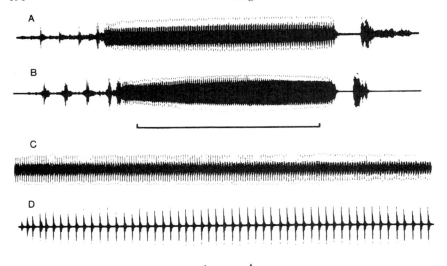

Figure 15.2. Oscillograms of typical signals of males (A and B) and females (C and D) of
N. lugens from rice-feeding populations (A and C) and *Leersia*-feeding populations
(B and D) from the Philippines. Time marks 1 s (after Claridge *et al.*, 1985b).

hybrids were obtained in the laboratory when no choice of mating was
provided.

Analyses of courtship signals showed significant differences between male
and, most obviously female courtship signals (Fig. 15.2). Laboratory hybrids
produced calls intermediate in pulse repetition frequency between the
parentals. We have recorded many field samples of these insects and have
never found evidence of intermediate calls and therefore of hybridization.
The two host-associated populations commonly occur in very close proximity
and thus must be regarded as truly sympatric. We must therefore conclude
that these very closely related forms morphologically attributable to *N. lugens*
really represent two distinct biological species. On the other hand, Saxena
and Barrion (1985) regard the *Leersia* associated populations to be a 'non-
virulent biotype' of *N. lugens*. They thus regard the differences between these
two biological species as no greater than those between the IRRI biotypes!

These long-term studies on *N. lugens* illustrate graphically the sort of
problems that are faced if we are to understand the diversity of any insect
pests. The arbitrary use of terms such as biotype, without adequate investigation, may simply obscure major variation and provide falsely simple
solutions to complex problems.

Pest community diversity

It is now widely recognized that an insect pest forms just part of a very
complex ecosystem with interactions usually with many other species of

organisms. Obviously the host-plant, the crop, is one of these, as discussed above. However, additionally the pest itself will almost always be just one of a number of herbivores in the system and all of these will also be attacked by a number of natural enemies, both predators and parasitoids. The role of natural enemies is central to the biological control of insect pests. The diversity of natural enemies of insect pests is usually very high and almost certainly generally responsible for the natural control of insect herbivores in most non-agricultural systems. The disruption of natural enemy complexes by the indiscriminate application of broad-spectrum insecticides has been a major cause of many pest outbreaks. For example, it seems likely that the brown planthopper, discussed above, has only become an important pest of rice in tropical countries because of the dramatic reduction of natural enemy numbers by pesticide applications (Kenmore *et al.*, 1984). Indeed, pest resurgence can commonly be induced by injudicious insecticide application.

Very large numbers of natural enemies of pests are parasitoids. These are very specialized predators that mostly belong to the two very large insect orders Hymenoptera and Diptera. In general, the basic morphological taxonomy of these insects is even less well-known than for the major pest groups themselves. Most detailed studies of these insects demonstrate that biological species are not adequately elucidated by morphological markers alone. Examples abound of failures in biological control due to inadequate taxonomy. There is an urgent need for basic taxonomic research on these groups together with biosystematic studies on major genera associated with important pests. For example, recent work on the important egg parasitoids of *N. lugens* has shown that what was previously thought to be one species of *Anagrus*, *A. flaveolus*, includes at least six biological species with particular host preferences and behaviour (Claridge *et al.*, 1987). Such studies urgently need to be expanded to natural enemies of other important pests.

Conclusions

A major problem in all areas of sustainable agriculture is the lack of basic research on insect pest and natural enemy taxonomy. The problem is undoubtedly greatest in tropical countries where the needs are in turn greater. There is an urgent requirement to pool expertise to provide this vital information. A system of networking, such as that proposed by Haskell and Morgan (1988), would be most appropriate. CAB International is ideally suited to co-ordinate such a system.

Given the basic taxonomic framework, it is essential to identify the major pest problems of the tropical world and then to target them urgently for detailed biosystematic studies. These will require multidisciplinary approaches that are well-suited to collaborative work between developed and developing world laboratories.

Finally, there is an urgent need to provide training for taxonomists at all levels. The lack of funding and support for taxonomic work in recent years has in turn lead to a reduction in training. If funding were available, universities, and other relevant institutes would be able rapidly to increase numbers of research studentships and relevant postgraduate courses.

If the world is serious about its interests in sustainable agriculture and the conservation of biological diversity, then high priority must be given to targeted funding of taxonomic research and training programmes.

Acknowledgements

I am deeply indebted to my many colleagues over the years in Cardiff who have contributed to our work on rice pests and to the Natural Resources Institute (NRI) for long-term financial support.

References

Blackman, R.L., Brown, P.A., Furk, C., Seccombe, A.D. and Watson, G.W. (1989) Enzyme differences within species groups containing pest aphids. In: Loxdale, H.D. and den Hollander, J. (eds), *Electrophoretic Studies on Agricultural Pests*. Clarendon Press, Oxford, pp. 271–295.

Cain, A.J. (1954) *Animal Species and their Evolution*. Hutchinson, London.

Claridge, M.F. (1985) Acoustic behaviour of leafhoppers and planthoppers: species problems and speciation. In: Nault, L.R. and Rodriguez, J.G. (eds), *The Leafhoppers and Planthoppers*. John Wiley, New York, pp. 103–125.

Claridge, M.F. (1987) Insect assemblages – diversity, organization, and evolution. In: Gee, J.H.R. and Giller, P.S. (eds), *Organization of Communities Past and Present*. Blackwell Scientific Publications, Oxford, pp. 141–162.

Claridge, M.F. (1988) Species concepts and speciation in parasites. In: Hawksworth, D.L. (ed.), *Prospects in Systematics*. Clarendon Press, Oxford, pp. 92–111.

Claridge, M.F. (1990a) Variation in pest and natural enemy populations – relevance to brown planthopper control strategies. In: Grayson, B.T., Green, M.B. and Copping, L.G. (eds), *Pest Management in Rice*. Elsevier, London, pp. 143–154.

Claridge, M.F. (1990b) Acoustic recognition signals: barriers to hybridization in Homoptera *Auchenorrhyncha*. *Canadian Journal of Zoology* 68, 1741–1746.

Claridge, M.F., Claridge, L.C. and Morgan, J.C. (1987) *Anagrus* egg parasitoids of rice-feeding planthoppers. In: Vidano, C. and Arzone, A. (eds), *Proceedings of 6th Auchenorrhyncha Meeting, Turin*. Institute of Agricultural Entomology and Apiculture, University of Turin, Turin, pp. 617–621.

Claridge, M.F. and den Hollander, J. (1982) Virulence to rice cultivars and selection for virulence in populations of the brown planthopper *Nilaparvata lugens*. *Entomologia Experimentalis et Applicata* 32, 213–221.

Claridge, M.F. and den Hollander, J. (1983) The biotype concept and its application to insect pests of agriculture. *Crop Protection* 2, 85–95.

Claridge, M.F., den Hollander, J. and Furet, I. (1982) Adaptations of brown plant-hopper (*Nilaparvata lugens*) populations to rice varieties in Sri Lanka. *Entomologia Experimentalis et Applicata* 32, 222–226.

Claridge, M.F., den Hollander, J. and Haslam, D. (1984*a*) The significance of morphometric and fecundity differences between the 'biotypes' of the brown planthopper, *Nilaparvata lugens*. *Entomologia Experimentalis et Applicata* 36, 107–114.

Claridge, M.F., den Hollander, J. and Morgan, J.C. (1984*b*) Specificity of acoustic signals and mate choice in the brown planthopper *Nilaparvata lugens*. *Entomologia Experimentalis et Applicata* 35, 221–226.

Claridge, M.F., den Hollander, J. and Morgan, J.C. (1985*a*) Variation in courtship signals and hybridization between geographically definable populations of the rice brown planthopper, *Nilaparvata lugens* (Stål). *Biological Journal of the Linnean Society* 24, 35–49.

Claridge, M.F., den Hollander, J. and Morgan, J.C. (1985*b*) The status of weed-associated populations of the brown planthopper, *Nilaparvata lugens* (Stål) – host race or biological species? *Zoological Journal of the Linnean Society* 84, 77–90.

Claridge, M.F. and Morgan, J.C. (1987) The brown planthopper, *Nilaparvata lugens* (Stål), and some related species: a biotaxonomic approach. In: Wilson, M.R. and Nault, L.R. (eds), *Proceedings of the 2nd International Workshop on Leafhoppers and Planthoppers of Economic Importance*. CAB International Institute of Entomology, London, pp. 19–32.

den Hollander, J. and Pathak, P.K. (1981) The genetics of the 'biotypes' of the rice brown planthopper, *Nilaparvata lugens*. *Entomologia Experimentalis et Applicata* 22, 76–86.

Diehl, S.R. and Bush, G.L. (1984) An evolutionary and applied perspective of insect biotypes. *Annual Review of Entomology* 29, 471–504.

Dobzhansky, T. (1937) *Genetics and the Origin of Species*. Columbia University Press, New York.

Everson, E.H. and Gallun, R.L. (1980) Breeding approaches in wheat. In: Maxwell, F.G. and Jennings, P.R. (eds), *Breeding Plants Resistant to Insects*. John Wiley, New York, pp. 513–533.

Haskell, P.T. and Morgan, P.J. (1988) User needs in systematics and obstacles to their fulfilment. In: Hawksworth, D.L. (ed.), *Prospects in Systematics*. Clarendon Press, Oxford, pp. 399–413.

Heinrichs, E.A., Medrano, F.G. and Rapusas, H.R. (1985) *Genetic Evaluation for Insect Resistance in Rice*. International Rice Research Institute, Los Banos.

International Rice Research Institute (1979) *Brown Planthopper: Threat to Rice Production in Asia*. International Rice Research Institute, Los Banos.

Kenmore, P., Carino, F.O., Perez, C.A., Dyck, A. and Guttierez, A.P. (1984) Population regulation of the rice brown planthopper (*Nilaparvata lugens* Stål) within rice fields in the Philippines. *Journal of Plant Protection in the Tropics* 1, 19–37.

Loxdale, H.D. and den Hollander, J. [eds] (1989) *Electrophoretic Studies on Agricultural Pests*. Clarendon Press, Oxford.

Mayr, E. (1942) *Systematics and the Origin of Species from the Viewpoint of a Zoologist*. Columbia University Press, New York.

Mayr, E. (1963) *Animal Species and Evolution*. Oxford University Press, Oxford.

Menken, S.B.J. (1989) Electrophoretic studies on geographic populations, host races, and sibling species of insect pests. In: Loxdale, H.D. and den Hollander, J. (eds), *Electrophoretic Studies on Agricultural Pests.* Clarendon Press, Oxford, pp. 181–202.

Menken, S.B.J. and Ulenberg, S.A. (1987) Biochemical characters in agricultural entomology. *Agricultural Zoology Reviews* 2, 305–360.

Paterson, H.E.H. (1985) The recognition concept of species. In: Vrba, E.S. (ed.), *Species and Speciation.* [Transvaal Museum Monograph No. 4.] Transvaal Museum, Pretoria, pp. 21–29.

Saxena, R.C. and Barrion, A.A. (1985) Biotypes of the brown planthopper *Nilaparvata lugens* (Stål) and strategies in deployment of host plant resistance. *Insect Science and its Application* 6, 271–289.

Saxena, R.C. and Barrion, A.A. (1987) Biotypes of insect pests of agricultural crops. *Insect Science and its Application* 8, 453–458.

Saxena, R.C. and Khan, Z.R. (1989) Factors affecting resistance of rice varieties to planthopper and leafhopper pests. *Agricultural Zoology Reviews* 3, 97–132.

Saxena, R.C. and Rueda, L.M. (1982) Morphological variations among three biotypes of the brown planthopper *Nilaparvata lugens* in the Philippines. *Insect Science and its Application* 3, 193–210.

Seshu, D.V. and Kauffman, H.E. (1980) Differential response of rice varieties to the brown planthopper in international screening tests. *International Rice Research Institute Research Paper Series* 52, 1–13.

Sixteen

General Discussion: Session III

Session Chairman: G.H.L. Rothschild, *Australian Centre for International Agricultural Research, GPO Box 1571, Canberra ACT 2601, Australia*

Session Rapporteur: J.D. Holloway, *International Institute of Entomology, 56 Queen's Gate, London SW7 5JR, UK*

Introduction

In this session the focus has been on the role of biodiversity in the management of pests. In this regard it is important to consider pests in the broad sense, covering not only insects but plant diseases, weeds, and diseases of animals. While the emphasis has been on insects, the management implications for other groups of pest organisms need to be noted.

We have to consider how biodiversity relates to pest management, and to see how management practices affect biodiversity. Further, we have to recognize that biodiversity *per se* does not necessarily confer stability to the system being considered. The question is how diverse a system needs to be in order to have a sustainable system.

Discussion

Lal: When we talk about low-input sustainable agriculture we have to be careful. The concept may be different in North America as opposed to resource-based agriculture in West Africa. It must be recognized that land exploitation without inputs can also be hazardous to both the resource base and the environment.

Altieri: I believe it is also important to understand why systems become unsustainable. This is often due not to population but to land concentration and access to it. The reasons are not only biological; social and economic forces are crucial, and in South Africa it is such constraints that lead to land being abandoned.

Siddiqi: With regard to mixed cropping and maintaining weeds in cultivated land, there is a danger from plant–parasitic nematodes, many genera of

The Biodiversity of Microorganisms and Invertebrates: Its Role in Sustainable Agriculture.
Edited by D.L. Hawksworth. © CAB International 1991.

which are ectoparasites and polyphagous. Their populations can be maintained on weeds. When crops are absent in post-harvest periods, if such weeds remain, this provides a means of maintaining and promoting harmful nematode populations which will ultimately be detrimental to crop production.

Hadley: Dr Altieri (Chapter 14) drew attention to two unsustainable systems of agriculture: (i) high-input, high-yielding simplified systems in North America where the increasing gross income of farmers is paralleled by stagnating or declining net incomes to farmers; and (ii) pastures that have replaced tropical forests in parts of Latin America, fuelled by government incentives, but where yields have quickly fallen off and sustainability has proved illusory. Can we be sanguine about the prospects of mixed diversified agricultural systems providing one way forward in making these two 'failed' systems more sustainable, and what subsidies or incentives might be given?

Altieri: Local agriculture is led by international forces, for example export demand in the case of North America. Policy changes will be necessary and taxes will need to be modified to provide incentives for sustainable agriculture. With regard to Dr Siddiqi's comment on nematodes, I concur that all biodiversity is not necessarily good, but we also have to remember that there are mixed systems developed by farmers and non-governmental organizations in Latin America which nevertheless appear to be working although the reasons for that are not understood.

R. Baker: Is it to be understood that low-input systems as practised in developing countries, and which can be seen as success stories, are only applicable to those countries or do they also have a place in developed countries – especially where the environment is under threat from the excessive use of chemicals?

Altieri: In their particular designs, such systems only fit the socio-economic and cultural reality of the Third World. However, we can derive principles from them that govern the sustainability of those systems and apply those ecological principles in the design of sustainable agro-ecosystems for industrial countries.

Jeger: With respect to scientists 'scouring the tropics' for possible biocontrol agents, these are often taken to developed countries for exploitation and may be subject to intellectual property rights. What can or should be done to protect the rights of countries from whence the biocontrol agent came?

Waage: You have identified a serious problem. It is reasonable for countries to view their indigenous pathogens as resources, as much as their oil production, and for them to protect their exploration – as already happens in Brazil. However, present patent laws protect the discoverer or describer, and not the country in which something is discovered. There is the potential for this to lead to problems, especially in the exploitation of tropical diversity by multinationals. At the same time, countries could help each other by exchanging agents and technologies; there is much potential for south–south exchange.

Robinson: Dr Waage has spoken of his concern that countries of origin of organisms that provide the raw material for biotechnological development by the developed countries will rarely if ever receive their full due. The application of the principle of intellectual property rights to genetically unaltered as well as 'engineered' strains of organisms could become a political and legal minefield. However, there is a danger that studies of biodiversity, far removed from biotechnology, may seriously be impeded by the hurried emplacement of legislation designed to protect commercial interests. Unnecessarily draconian restrictions and prohibitions on the movement of specimens, samples, and information may create grave bureaucratic problems in the process of 'taxonomic technology transfer'.

Waller: Some biocontrol of plant pathogens occurs naturally in many agricultural systems, and is associated with a diversity of microorganisms on plant surfaces, especially on perennial crops, and in soil. These natural systems can be disturbed by injudicious use of fungicides which leads to greater incidence, for example in relation to coffee berry disease in Kenya. We know too little about the functions of the diverse organisms involved to manipulate the system to advantage. There is a need to better understand the functional diversity of organisms involved in pathogen inhibition. Diversity among pathogens is negatively important as such variability represents the source from which 'new' pathogens or more virulent strains emerge (particularly through selection pressure imposed by mass efforts to improve crop yields by plant breeding and monoculture). The need is to reduce the diversity among pathogens by first understanding and then manipulating the mechanisms which sustain it, or at least to segregate its effects by matching it with host diversity, thus reducing selection pressure and so sustaining the durability of resistance.

Greenland: Differentiation at the infraspecific level introduces an additional dimension to the problems of what biodiversity is desirable, and what needs to be done about it. Do we need to have live insects and acoustic laboratories in our taxonomic institutes? It appears at this stage that we may be able to say little more than that we need to do much more research on the infraspecific level of differentiation, including the biochemical characterization of both invertebrates and microorganisms.

Perfect: Interaction between pests in crops is another problem in pest management; approaches have too often been unimodal. A control intervention for one species is also frequently the determinant of the population of another. Less biodiversity in the pest fauna can therefore be favourable. Decision-making is hindered through lack of knowledge of pest interactions, for example in the context of economic thresholds.

Kishimba: It has been correctly pointed out that pesticides may wipe out some of the useful microorganisms and insects, but in tropical countries fire is one of the tools used to generate fresh palatable grasses. It would be of interest to know the effects of fire in maintaining biodiversity.

Altieri: Many ecosystems have evolved with fire, and fire-adapted communities are dependent on fire for regeneration. Prescribed burning can be used effectively to maintain and enhance biodiversity in such systems, including grasslands. However, fire can also be used wisely in fallow systems with beneficial effects.

Kishimba: It seems to me that biodiversity is bound to improve sustainable agriculture, but the contrary view has also emerged. Should we be encouraging succession in pest management as a way of maintaining stability?

Altieri: A sustainable agricultural system is one capable of being self-sufficient by sponsoring its own soil fertility, crop protection, and crop productivity. No chemical inputs are used, but instead biodiversity is employed to maintain the functions of soil fertility, pest regulation, yield, etc. Mimicking natural succession with analogous cropping systems simulating successional features such as plant diversity or architecture tends to improve pest control.

Dhiman: In this session attention has been drawn to a wide range of interesting but very diverse topics. A great deal of information useful to our understanding and maintaining of biodiversity is evidently present, but scattered and fragmentary. In order to facilitate some practical output, the information from diverse interdisciplinary sources needs to be consolidated and integrated. I see a need for an interdisciplinary group to compile the available information and data on various facets of biodiversity so that further headway can be made without having to start afresh.

Session Chairman's conclusions

Summary

Biodiversity is a complex subject, even from defining the units that we have to consider whether at the species level or below. Determining the units will determine how complex a system is. I believe we will have to work at the reproductively isolated population level. With regard to funding research, the question is how much of the system do we actually need to know to achieve a practical outcome. There is the potential for looking at variation almost indefinitely which might be worthwhile as basic research and could produce spin-offs, but in practice judgements have to be made as to what are the key elements of the populations that we are dealing with. In order to do that we need to understand the functional relationships. Looking at the system is important, but even if that is done intuitively initially, crude simulation models to indicate what might be key variables in response to perturbations would be valuable. If a landscape or agricultural system is being modified, these are some of the issues which need to be addressed. There is a great need for further research, in extending the results of what is already known, and also on folklore which works but where the reasons why require investigation. Two particular threads emerge. First, there is clearly more need for work on biosystematics, but perhaps trying to focus on key elements. Second, there is a need to draw paradigms out of traditional farming systems, to provide guidelines for the setting up experimental networks whereby manipulative studies can be conducted. The research will have to be packaged in a form that is attractive to funding agencies.

Conclusions

This session was concerned with biodiversity and the management of pests. A need to define 'biodiversity' was recognized, despite the difficulties of that. In answer to the question as to why it was a need in pest management to seek diversity, the view was expressed that evidence for this was necessary and that the concept should not be accepted only on a theoretical basis.

If biodiversity was to be defined, it was suggested that the most appropriate units to be used might be the genome. The total characteristics of organisms and not just morphology were important. It follows that it is crucial to support biosystematics and to look at that subject from an interdisciplinary standpoint.

Given the objectives of the Workshop, more studies are clearly needed in relation to biodiversity and pest management. The question as to why it is important to study diversity will be asked. Evidence from experiments or case histories where greater diversity in agro-ecosystems results in better regulation of pests is necessary. In addition to biosystematics, a second priority is to study selected agro-ecosystems in depth to obtain such data. Once case histories were available, it was also suggested that it might be possible to simulate some of those systems on a smaller scale, manipulate them, and examine the effects of experimental perturbations. If this could be standardized, a network could be established so that parallel experiments on different groups would be available.

There is a need to examine the economic rationale for exploring biodiversity in relation to pest management. As pointed out by Professor Altieri (Chapter 14), if you are looking at introducing new systems such as mosaics, in the short term there are going to be economic questions, even though in the long term sustainability might be achieved. The short term *versus* long term conflict requires the involvement of socio-economists.

In summary, biosystematics and inventory work, and case histories and experimental manipulation of environments, were identified as necessary to provide a solid scientific basis for understanding biodiversity in pest management.

IV
Biotechnology and Biodiversity among Invertebrates and Microorganisms

Seventeen

Biotechnology and Biodiversity

A.T. Bull, *The International Institute of Biotechnology, Biological Laboratory, University of Kent, Canterbury, Kent CT2 7NJ, UK*

ABSTRACT: Biotechnology has been dominated by molecular biology, but the importance of biodiversity is becoming increasingly clear through the demand for both novel products and microorganisms that can be deployed in the environment. Remarkably little is known of microbial diversity, numerous species remain to be described, and the genetic diversity within those that are known is scarcely studied. Many microorganisms have yet to be cultured. The priorities in microbial diversity studies are to develop new isolation and cultivation methods, and to detect and make accessible non-culturable microorganisms by molecular methods. Conservation of the undiscovered potential is best effected by the conservation of ecosystems; culture collections can never hold more than a fraction of the resource. Taxonomy and molecular biology can assist in the search for novel microorganisms, but targeted screening and other methods are now the focus of attention. While global actions on the erosion of biodiversity are required, a bilateral programme 'Biodiversity for Innovation' is investigating mycorrhizal inoculants and biodegraders in Indonesia; this approach provides a way of assisting in both the conservation and exploitation of the tropical microbial resource.

Microbial diversity

Why, in the context of biotechnology, should it be necessary to know about biodiversity, and to be concerned about its conservation? Biodiversity defines the variety and variability of all forms of life, the ecological complexes in

The Biodiversity of Microorganisms and Invertebrates: Its Role in Sustainable Agriculture.
Edited by D.L. Hawksworth. © CAB International 1991.

which they occur, and the ecological processes of which they are part; it comprises the diversity found at the genetic, species and ecosystem levels (National Science Board, 1989; McNeely *et al.*, 1990). Biotechnology in the past two decades has been dominated by the spectacular developments in molecular biology, but the counterbalance, available from biodiversity, has been advocated on a previous occasion (Bull, 1987), and the force of the argument strengthened during the intervening period. The unique opportunities provided by recombinant DNA technology are incontestable, but for many biotechnological products and processes this option is either unnecessary, inappropriate, or even proscribed. The demand for totally novel or for replacement products on the one hand, and the deployment of microorganisms in the environment on the other (Bull, 1991) exemplify the need for biotechnologies based upon natural organisms. Moreover, the advent of genetic engineering highlights the importance of natural gene pools for purposes of improving crop plants and domestic animals, and the development of industrial microorganisms. Although the emphasis will be placed on microbial diversity in this contribution, it is ironic that, despite the known commercial value of microorganisms, so little is known of the extent of their diversity or of the complexity of their roles in sustaining global 'life-support systems' (World Conservation Strategy, 1980), such as agricultural, watershed forest, coastal, and freshwater ecosystems. Thus, a *prima facie* case for conservation (the planned management of the biosphere for the greatest sustainable benefit of the present while preserving its potential for the future) can be upheld on the grounds of preserving and utilizing microbial diversity.

The knowledge base

What is known about microbial diversity?

The answer to this question is: remarkably little. The published inventories (e.g. Wison, 1988) put the aggregated number of bacteria, fungi, and protozoa at 82 500, but we can only guess at the total number of microbial species. Granted that even within particular kingdoms the reported species numbers may vary disconcertingly (cf. Wilson, 1988*a*; Hawksworth, 1991; Hawksworth and Mound, Chapter 3); the comment that such numbers probably represent much less than 10% of the real fungal flora of the world alone is salutary. The emphasis of many investigations in microbiology has been on the functional characteristics of microorganisms, either because of recalcitrant taxonomy or a disdain for biosystematics. The latter, traditionally, is viewed as the means for assessing biodiversity and consequently past efforts in microbial ecology have tended to concentrate on particular organisms, or on functional properties of the community or ecosystem. In contrast, there is very little understanding of the genetic diversity which exists within the population of a given species. Indeed, even the species concept in micro-biology, and singularly in bacteriology, needs to be applied with great caution because of the extent of horizontal gene flow which may take place in microbial populations. Perhaps, as Verstraete and Huysman (1988) have advised, it is necessary now 'to study the ecology of genes in the microbial community rather than the organisms they are packaged in'. The identity of a *nif* gene in the marine cyanobacterium *Trichodesmium* (which has not been cultured axenically) is a good case in point (Zehr and McReynolds, 1989).

Similarly, attempts to determine diversity indices of microorganisms in natural ecosystems are all too readily thwarted by our inability to culture (and hence 'identify') representative samples of such communities. It is often remarked that less than 10% of the bacteria in soils (as determined by microscopy) and less than 0.1% and as low as 0.001% in the open seas are, as yet, culturable. However, a realistic approach to the estimation of genetic diversity within microbial communities is now possible by adopting tech-niques of molecular biology, such as 16S rRNA and tRNA analysis. Whatever the remaining difficulties of defining genetic diversity, including the con-ditions of dormancy and regulation in the environment (Bock, 1989), the awareness of such diversity is essential for both effective natural resource management and as the source of novel microorganisms (or genes) for biotechnological innovation.

What are the priorities for microbial diversity studies?

The US National Science Board (1989) recommended that the following actions were required in order to provide 'an acceptable standard of know-ledge about the world's biota': (i) acquisition of a complete global inventory; (ii) comprehensive research of representative and threatened global locations, including those with a high degree of endemism; (iii) conservation; (iv) development of comprehensive databases; (v) development of human resources; and (vi) fostering international programmes. Whether attempts to describe the totality of microbial diversity are practicable given the problems of gene flux and recognition referred to above is debatable. Unquestionably, what is required for biotechnology purposes is knowledge of the range of microbial diversity and the means for its discovery and recovery from the environment.

Concerning methodologies

In general terms, there is a need to develop new isolation and enrichment cultivation methods in order to obtain novel organisms. Clearly, the choices of the ecosystem or habitat to be sampled has a bearing on the outcome of such searches. There is an implicit feeling that a focus on exotic habitats is likely to reveal the greatest novelty, but that this is not always the case is well demonstrated by the discovery of magnetotactic bacteria in 1975 from temperate freshwater ecosystems (Blakemore, 1975). In other cases, it is the presence of macroorganisms in a particular ecosystem that foretells the probable existence of novel microorganisms [vide the bacterial basis of life in deep-sea hydrothermal vents (Jannasch, 1989) and shallow-sea petroleum seeps (Alper, 1990)]. Much has been written by microbial ecologists on the problems of sampling ecosystems; the procedures adopted are often intuitive, but, at the very least, it is important to remember that a given environmental sample represents a mixture of micro-habitats or ecological niches, and that its investigation should take cognisance of this fact. Baiting an ecosystem prior to sampling can be a productive way of searching for organisms having particular functional characteristics. In a comparable manner, the pretreatment of samples prior to their microbiological analysis may elicit the selective recovery of particular microbial populations. For example, the exposure of samples to heat, antimetabolites, or bacteriophages, or deploying them in enrichment procedures, will bring about a preliminary discrimination of the microflora on functional grounds. Whatever means of sampling and recovery are used, the investigator can never be certain what proportion of the total microbial population has been revealed. However, direct examination of the sample, say by epifluorescence microscopy, assuredly will show that recovery invariably is poor. Enrichment techniques are often used in this context: they are designed to alter the environment such that organisms of interest should out compete all others in the sample and hence become the dominant population. The deployment of chemostat culture for such isolation frequently enriches novel bacteria (as judged by the failure of identification services to assign them to known species or genera), even from supposedly well-characterized environments. The inherent properties of chemostat cultures provided for the establishment of unique enrichment conditions (e.g. with respect to specific growth rate, the nature and concentration of growth rate limiting substrate, the concentration of organisms, and environmental conditions) which, in turn, can enable the dissection of populations in terms of their physiological properties. The most effective use of chemostat (and other) enrichment isolation methods necessitates an appreciation of microbial interactions (especially obligatory ones; Bull and Slater, 1982; Parkes, 1982). It may seem self-evident that a knowledge of the physico-chemical and macrobiotic characteristics of the ecosystem will be critical when defining culture media and incubation conditions designed to isolate microorganisms from it, but this point warrants restatement. Similarly, culture media that have been formulated to elicit the maximum isolation of microorganisms (e.g. Hütter, 1982), usually only serve as a crude guide for individual studies and ultimately the investigator is forced to define his or her particular requirements.

A second cluster of problems surrounds the characterization of non-culturable microorganisms: how can they be detected in the environment, distinguished, identified and made accessible for biotechnology? The application of DNA restriction endonuclease analysis, rRNA sequence analysis, and DNA probes, has revolutionized our ability to address such questions. Thus, the restriction endonuclease analysis of microbial genomes is a valuable means of differentiating bacteria at the level of genus, species, strains and mutants, and it is reliable and readily applied to large numbers of isolates and to biomass taken directly from the environment. An early application of the technique enabled Mielenz *et al.* (1979) to correct claims that *Rhizobium trifolii* nodulated soyabean, and to show that such reported nodulation was due to *Rhizobium japonicum*. Numerical analysis of DNA fingerprints not only facilitates pattern analysis as a basis of classification, but also may reveal correlations between the fingerprint and phenotypic attributes, such as antibiotic resistance, as in *Neisseria* (Sorensen *et al.*, 1985). The choice of endonuclease is important for achieving the unambiguous differentiation of strains, and the use of 'rare' restriction enzymes may be the preferred option. A recent analysis of strains of *Azospirillum* (Giovanetti *et al.*, 1990) exemplifies this point.

The use of ribosomal RNA sequences in defining the components of mixed, natural microbial populations has been pioneered by Pace *et al.* (1986). Ribosomal RNAs have highly conserved structures and nucleotide sequences, and are well-suited for investigating phylogenetic relationships; to date 5S and 16S rRNA molecules have been used in this context. The 5S rRNAs are comparatively small molecules (approximately 120 nucleotides), and the sequences of several hundred are now known. Sufficient rRNA for the analysis can be provided by about 10^9 bacteria, but Rogall *et al.* (1990) have shown that as few as 100 bacteria are sufficient if the rRNA is reverse-transcribed into cDNA and then amplified by the polymerase chain reaction (PCR). The caution against sampling bias mentioned above also applies to the collection of biomass for these analyses. The practical considerations of RNA isolation, sequence determination, and sequence alignment are introduced by Pace *et al.* (1986), who also present the results of analysing several mixed populations; including a thermal pool in Yellowstone National Park, Octopus Spring, hypothermal vent symbionts, and copper-leaching operations in New Mexico.

The analysis of natural populations by cloned 16S rRNA sequences is superior to 5S rRNA for several reasons, most particularly because of its greater information content, which enables greater precision in the establishment of phylogenetic relationships. This technique also has been applied to the analysis of bacterial populations in Octopus Spring (Ward *et al.*, 1990). Eight different bacteria were detected by this means, none of which have ever been cultured from this ecosystem, whereas earlier examination of this ecosystem by 5S rRNA analysis revealed the presence only of three organisms, two eubacteria (probably *Thermus* species) and a sulphur archaebacterium (Pace *et al.*, 1986). The question of whether it might be possible to correlate physiological attributes with nucleotide fingerprints in a manner analogous to nucleotide hybridization probes, has obvious relevance to the detection of biochemical novelty. And, likewise, the great challenge to microbiologists is to develop means of cultivating as yet non-cultured organisms.

Conservation

The crisis surrounding the threat to biodiversity is being articulated and documented with increasing regularity and purpose (Wilson, 1988b; McNeely et al., 1990). This crisis has its origins in several interrelated acts of environmental mismanagement, most notably: chemical pollution, over-exploitation of natural renewable and non-renewable resources, large-scale physical interventions resulting in land- and water-use changes; and popu-lation growth. The magnitude of the threat has been quantified with respect to animal and plant species, and to habitats (McNeely et al., 1990), but published figures almost certainly underestimate the rates of habitat destruc-tion. Tropical forests, wetlands, and coral reefs, are ecosystems of extreme diversity and consequently excite particular concern. Consequences of the loss of genetic diversity in animals and plants for biotechnology are relatively easy to estimate. In broad terms only about 1.5% (4 000) of flowering plants have been screened for pharmaceutical compounds, and of this number about 3% currently provide major drugs. Thus, the undiscovered potential is very considerable. Set against these statistics are those on the predicted rate of plant species loss: average tropical deforestation of 0.6% per annum, with about 25% of plant species associated with forests likely to be lost within 20 years (Raven, 1988).

Predictions of the loss of microbial diversity are very much more prob-lematical due to the meagre knowledge base and the evidence for loss tends to be anecdotal. That loss does occur is undisputed. Rifai (1989) reports the loss of *Penicilliopsis clavariaeformis* from the Bogor Botanic Gardens because the appropriate *Diospyros* tree species with which it is associated is no longer available, and the virtual disappearance of *Cookeina tricholoma* from Java as a consequence of the island's environmental disturbance. Many microorganisms may be resistant to upheaval of the environment, but others, particularly those located in pristine ecosystems and intimately associated with microorganisms, may be less ubiquitous, and thereby susceptible to habitat perturbation or destruction.

Traditionally microorganisms have been maintained in culture collections or herbaria. However, I am convinced that *in situ* conservation is as crucial for the protection of microbial genetic diversity as it is for the protection of macroorganisms and 'conserving biological diversity equals conserving ecosystems' (International Union for the Conservation of Nature and Natural Resources, 1989) is a message with equal validity for microbiologists and biotechnologists as it is for botanists and zoologists. The World

Conservation Strategy (International Union for the Conservation of Nature and Natural Resources, 1980) was only partly right when it spoke in terms of 'isolating from nature a (microbial) strain with a particular desired property is tedious' and 'such strains need to be maintained in culture collections'. Without question culture collections fulfil several vital requirements for biotechnology, but they can never expect to hold more than a small fraction of the genetic diversity found in nature, or to guarantee against the loss of desirable phenotypes while the organism is in store or culture. This is not the place to detail the procedures that are available for preserving and maintaining microbial strains. The choice of methods is large and can be made on the basis of anticipated storage time, ease of retrieval, and the sensitivity of the organism to damage. Increasingly, genetic diversity can be stored artificially as nucleic acid sequence data.

The discovery of novelty

Omura (1986), in a seminal paper on the philosophy of drug discovery, pointed to a number of factors which have general relevance to search and discovery activities: (i) belief in the great capabilities or microorganisms; (ii) development of well-designed screening systems against given objectives; (iii) recognition that screening is not a mere routine operation; (iv) emphasis on basic research, and (v) treasuring good human relations.

Screening operations require a multidisciplinary approach and depend upon: an ability to devise procedures for the isolation and cultivation of microorganisms, particularly culture conditions that lead to the expression of metabolic pathways that are to be exploited; the development of sensitive biological systems (whole organisms, whole cells, enzymes) for the detection of desired activity; and an ability to assay small quantities of substances. Microorganisms can be isolated from an enormously wide range of environments on the planet, including antarctic soils, saline lakes, hydrothermal vents. Also, because of their ubiquitous association with animals and plants, especially promising sources of microorganisms can be expected in environments with rich floras and faunas.

Current interest in exploring the potential of marine organisms as a new resource for biotechnology is illustrated by the substantial investment being made in Japan. A new Marine Biotechnology Institute, established in 1988 by a consortium of 24 private companies at an initial cost of about £250 million, is being made available for Japanese and overseas researchers. Research and development is to focus on technology for the utilization of marine organisms, and chemicals and biofunctions of marine origin. The organisms to be investigated will be microorganisms, algae (including seaweeds), sponges, and coelenterates. Exciting though the prospects are of discovering dramatically novel organisms, and thence biochemistry, by searching in exotic environments (e.g. members of the Archaea; Woese *et al.*, 1990), it has to be remembered that new taxa of microorganisms are regularly reported as isolates from even temperate agricultural soils and other commonplace habitats. Even such a banal environment as stale beer has recently revealed a new genus of bacteria, *Pectinatus* (Lee *et al.*, 1978). New genera and species of microorganisms are being reported constantly, and the following examples represent the results of a very cursory skim of the literature: *Actinopolyspora*, *Nitrobacterium* and *Nitronococcus* (haloakalophiles); *Pyrobaculum* and *Thermococcus* (hyperthermophiles); *Methanococcus jannaschii* (thermophilic methanogen); *Aquaspirillum* and *Bilophococcus* (magnetotactic bacteria); *Haloferax mediterranei* (halophilic PHB producer); *Thiobacillus cuprinus*; *Heliobacterium mobilis* (the first Gram positive nitrogen-fixing bacterium); *Culicinomyces bisporalis* (mosquitocidal); *Kibdelsporangium aridum* (peptide antibiotics); and *Desulfomonile tiedjei* (anaerobic, dehalogenating, sulphate-reducing). These few examples reveal the range of novel organisms that are continuously being isolated and described, and also certain of their properties which offer possibilities for biotechnological exploitation.

Search for novel microorganisms

The role of taxonomy

The important role of taxonomic methods in the search and discovery of new bioactive substances is increasingly becoming appreciated by industry as, also, is the need for greater numbers of trained microbial systematists (National Science Board, 1989). The case for investing in microbial systematics as an effective route to isolation, screening, and the recognition of novel taxa has been stated cogently by Goodfellow and O'Donnell (1989). These authors show that the availability of high fidelity microbial strain databases offer real opportunities for the design of media for selective isolation and the detection of novel or rare organisms. Omura (1986) strongly advocated taxonomic studies on strains shown to be producing new substances because other members or close relatives of that taxon similarly may yield novel exploitable properties. Thus, taxonomic studies within the new actinomycete genus *Kitasatosporia* revealed several species, including producers of a broad specificity herbicide (phosalacine), and an inhibitor of bacterial cell wall synthesis (setomimycin) (see Omura, 1986). The availability of computer programs provides a rational basis for formulating taxon-specific isolation media (e.g. DIACHAR; Sneath, 1980) or for identifying unknown strains using phenetic identification schemes (e.g. MATIDEN; Sneath, 1979). Goodfellow and O'Donnell (1989) caution that since the output from such programs is subject to several value judgements, their interpretation needs the skills of an experienced systematist.

The role of molecular biology

The impact of molecular biology techniques as a means of enquiring into the nature and extent of microbial diversity has been mentioned above. Here the potency of such techniques for revealing novelty is illustrated with reference to recent studies on the cyanobacterial community of Octopus Spring, Yellowstone National Park (Ward *et al.*, 1990). Using an improved 16Sr RNA sequence analysis, eight sequence types were detected that were distinct from those of any microorganism that had been cultured from this or similar geothermal habitats (p. 207). Thus, even this very well-studied microbial ecosystem contains many hitherto undetected and novel organisms. The implications of such findings are profound for ecologists, and also for biotechnologists who will want to screen the newly revealed genetic diversity for useful properties.

Searching for novel biotechnological properties

Targeted screening

The focus of modern search and discovery programmes is on the development of target-directed screens that produce high probabilities of success. Although such screens have been applied largely in the search for novel therapeutic agents, the same intelligent thinking can be adopted for other purposes. The following examples illustrate the ingenuity and success of this approach.

Mode of action screens

These screens were developed with conspicuous success in the search for novel antibiotics as a consequence of detailed knowledge of the chemistry and synthesis of bacterial cell walls. An elegant screen for bacterial cell wall synthesis inhibitors in tandem with antimycoplasma metabolites was described by Omura (1981), and based on: (i) the differential activity of culture supernatants against *Bacillus subtilis* and mycoplasma; and (ii) inhibition of the incorporation of diaminopimetic acid or leucine into the cell wall or proteins. This screen yielded several new antibiotics in addition to previously described compounds. Mode of action screens have been effectively exploited in the discovery of novel β-lactam antibiotics (Nisbet and Porter, 1989).

Enzyme inhibitors

One of the most significant innovations in recent times has been the screening for enzyme inhibitors of microbial origin (Umezawa, 1972; Schindler, 1980).

Such inhibitors have been developed or are in the process of being developed for hypertension (renin inhibitors), thromboses, obesity, stomach ulcers, and hypercholesterolaemia (hydroxyglutaryl coenzyme A reductase inhibitors).

Ligand mimicry

A current illustration of this tactic is that deployed in screens for gallium-accumulating bacteria that would function at high pH values and sequester Ga(III) under conditions of great excess of Al(III) (D.J. Gascoyne *et al.*, 1991). Organisms, selected under iron-limited conditions and in the presence of gallium, synthesized siderophores that had high affinity for Ga(III) and thus had the capacity for the selective accumulation of gallium.

Immunoregulation

The search for microbial metabolites that inhibit or stimulate the immune system has a large number of targets, and has been greatly encouraged by the discovery of cyclosporin A and its ability to retard organ transplant rejection. The lymphokine interleukin-1 (IL-1) elicits pleomorphic effects *in vivo* and is associated with acute inflammation and thence with the damage to joints (rheumatoid arthritis). Although cyclosporin A inhibits IL-1 production, there is now an intensive search for microbial IL-1 receptor antagonists with which to treat arthritic diseases.

Roles for genetics and molecular biology in target-directed screens

Genetics and molecular biology have had a remarkable enabling effect on screening by: (i) rendering test organisms more sensitive or resistant to known agents, for example antibiotic supersensitive strains in the search for novel β-lactams (Aoki *et al.*, 1976); (ii) *via* the development of molecular probes, for example DNA probes, monoclonal antibodies; and (iii) providing the opportunity for receptor screening *in vitro*, either as a receptor binding assay, or, as a functional assay which reflects the effect of binding. The success of this latter approach relies on the cloning of receptor genes into appropriate mammalian cell lines, for example, animal cells containing recombinant dopamine receptors as a means for target-directed screening for anti-psychotic and anti-anxiety metabolites (Netzer, 1990).

Automation

The development of target-directed screening has not eliminated the need to conduct very large numbers of assays. Netzer (1990) illustrates this with reference to Merck's receptor screens in which approximately 60 receptors are used to screen up to 40 000 microbial culture broths, and other chemicals, annually. Statistics such as these highlight the need for automated and often miniaturized screening operations. Automation may be possible at most stages of a search and discovery programme from the segregation of cells (e.g. flow cytometry), the selection of colonies after the plating out of cells (e.g. X–Y scanning), the production of culture broths for assay (e.g. microtitre plates), the assay (e.g. enzyme-linked immunosorbent assay), and to data processing (computation methods).

The use of parameter optimization

This strategy can be illustrated by the use made of discriminant analysis to provide positive selective pressures for the isolation of antibiotic-producing microorganisms. Discriminant analysis enables one to define characteristics which are important in distinguishing between groups of individuals within a population; it is dependent on finding suitable linear combinations of the original variables. One recent, novel application has been made by Huck (Bull *et al.*, 1990), who used discriminant analysis to identify nutritional and physiological parameters which successfully increased the competitiveness of antibiotic-producing strains on isolation plates. The optimization was applied in two stages: the first resulted in the definition of selective pressures which enhanced the competitiveness of actinomycetes *vis-a-vis* eubacteria and fungi, while the second greatly increased the competitiveness of antibiotic-producing actinomycetes *vis-a-vis* non-producing actinomycetes.

Serendipity and know-how

Despite the emphasis which has been afforded to intelligent search and discovery practices, it is worth restating the significance of chance observation, allied to serendipity, in the finding of novelty. Thus, the new extreme thermophile *Actinopolyspora halophila* (Gochnauer *et al.*, 1975) was discovered as a consequence of a chance infection of a batch of non-sterile culture medium intended for the cultivation of *Halobacterium* and left at room temperature for 50 days. Similarly, discovery of the new genus *Pectinatus* (Lee *et al.*, 1978; see p. 210) was the result of observing the contamination of beer that had been kept unintentionally for several weeks at 30°C.

Know-how reveals itself in the appropriate pre-treatment of samples prior to isolation. This may take the form of physical separation as in the flotation of spores away from the sample, as in the recovery of *Actinoplanes* from soil samples. The enrichment of communities of microorganisms, which may be obligatory in situations such as xenobiotic degradation, often leads to novel organisms being selected. Shelton and Tiedje (1984) isolated a bacterial community from sewage sludge that functioned co-operatively in converting 3-chlorobenzoate to methane. It was shown later (de Weerd *et al.*, 1990) that the organism responsible for the initial reductive dehalogenation reaction was a previously unknown genus and species of bacteria, *Desulfo-monile tiedjei.*

Case history: A UK–Indonesia co-operative programme

Here, I wish to refer to an international research and development programme which brings together most of the elements discussed earlier in this contribution. In July 1990, a bilateral governmental programme was launched in Indonesia under the title 'Biodiversity for Biotechnology Innovation' and involving, initially, academic research groups in the two countries. There were three chief reasons for the choice of Indonesia: (i) Indonesia's 'mega-diversity' status is exceeded only by Brazil and Colombia, and it possesses extraordinarily high endemism (McNeely *et al.*, 1990); (ii) conservation is a strategic issue in the Indonesian Government's Repelita V (current 5-year development plan); and (iii) mutual respect and confidence has built up over several years between the participating parties (the Inter-University Centres on Biotechnology in Bandung, Bogor, and Yogyakarta, in Indonesia, and The International Institute of Biotechnology in the UK), as a result of postgraduate research training, and a desire to establish long-term collaborative research activities.

The co-operative programme comprises two initial research components: (i) mycorrhizal inoculant technology for reafforestation by the development of natural mycorrhizal fungi to enhance the nutrition, growth and survival of tropical lowland plants in denuded soils; and (ii) biodegradation of toxic chemicals for environmental protection concentrating in the first instance on the detoxification of halogenated chemicals and the development of the 'best available technology not entailing excessive cost'. These development projects will be complemented by focused taxonomic assessments of the microbial diversity. The wider objectives of the project include: the improvement of Indonesian skills and expertise in environmental biotechnology research methods; and an increased awareness in Indonesia of the potential of careful management of the natural microbial gene pool for biotechnology innovation.

Mycorrhizal status is an important factor in tropical agriculture and in the regeneration of tropical rain forest, but the role of these fungi and of biological nitrogen fixers in sustaining soil fertility largely has been ignored. Two basic questions are being addressed as part of this programme. What is the diversity within Indonesia of mycorrhizal populations relative to other ecosystems, and are there novel or unusual fungi in Indonesian habitats that might have application elsewhere? And, how do indigenous mycorrhizal fungal populations change after forest clearance and replanting with the same or different tree species, or with other crop plants? We are interested in knowing if such changes can be manipulated so as to conserve tropical forests and their inherent biodiversity. The thrust of the detoxification project is to isolate and select novel microbial biocatalysts that can be deployed in the treatment of environmental halochemicals (e.g. chlorinated solvents, chloro-aromatics). It is known that certain white-rot fungi (lignin degraders) can transform complex chlorinated hydrocarbons and tropical forests offer unique opportunities to isolate novel wood-rotting fungi.

Marine ecosystems also are endowed with a spectacular biodiversity, and coral reefs in particular have a biodiversity and fragility that is comparable with that of tropical forests. Indonesia has the most extensive reefs in the Indo-Pacific Ocean and the greatest diversity of hard corals (McNeely *et al.*, 1990). A subsequent objective of The International Institute of Biotechnology is the establishment of a marine biotechnology programme, primarily focused on coral reef ecosystems. The development of a mainstream biotech-nology programme of this type also generates a spin-off into macrobiological diversity. In particular, molecular biological methods will be applied in the management of genetic diversity of rare tropical animals in a joint activity with the Durrell Institute of Conservation and Ecology.

The Agreement drawn up between the co-operating parties is in the spirit of the Biodiversity Convention (International Union for the Conservation of Nature and Natural Resources, 1988) with regard to Articles 14 and 15 on the access to wild specimens and payments for the use of biomaterials. Thus, all biological materials and information arising from the co-operative pro-gramme are freely available for research purposes to both parties, and any commercial exploitation will be made jointly.

Conclusions

It is my hope that this synoptic discussion has demonstrated something of the mutual interdependency of biotechnology and biodiversity. The resource that biodiversity brings to technological development literally is incalculable when one considers microorganisms. The prospect of an irrevocable loss of biodiversity should be as unacceptable to microbiologists and biotechnol-ogists as it is to botanists and zoologists. This discussion has centred largely on the industrial and health-care sectors of biotechnology, but those relating to agricultural production, soil conservation and fertility, biological contol, food processing, and the provision of clean water are no less important or less dependent on biodiversity. Undoubtedly the erosion of biodiversity is a problem of global proportions, and international actions are required for its solution and to promote the conservation of genetic diversity for sustainable biotechnologies. Given the disproportionate occurrence of biological diversity in the countries of the Southern Hemisphere, the question of 'biotechnologi-cal innovation for whom?' must be addressed. Modest bilateral programmes of the sort described above suggest one way forward while the National Science Board (1989) rightly emphasizes the necessary contributions required through training projects, research and development support, and the establishment and management of reserves.

Acknowledgements

I wish to thank the Overseas Development Administration for a contract in support of the Biodiversity for Biotechnology Innovation programme, and the many friends and officials in Indonesia who have encouraged its planning and implementation.

References

Alper, J. (1990) Oases in the oceanic desert. *American Society for Microbiology News* 56, 536–538.

Aoki, H., Sakai, H., Kohsaka, M., Konomi, T., Hosoda, J., Kubochi, Y., Iguchi, E. and Imanaka, H. (1976) Nocardicin A; a new monocyclic β-lactam antibiotic. I. Discovery, isolation and characterization. *Journal of Antibiotics* 29, 492–500.

Blakemore, R.P. (1976) Magnetotactic bacteria. *Science* 190, 377–379.

Brock, T.D. (1989) The study of microorganisms *in situ*: progress and problems. In: Fletcher, M., Gray, T.R.G. and Jones, J.G. (eds), *Ecology of Microbial Communities*. Cambridge University Press, Cambridge, pp. 1–17.

Bull, A.T. (1987) The biotechnologies of the nineties. In: Neijssel, O.M., van der Meer, R.R. and Luyben, K.C. (eds), *Proceedings, 4th European Congress on Biotechnology*. Elsevier, Amsterdam, vol. 4, pp. 189–202.

Bull, A.T. (1991) Degradation of hazardous wastes. *Philosophical Transactions of the Royal Society, B*, in press.

Bull, A.T. and Slater, J.H. (1982) Microbial interactions and community structure. In: Bull, A.T. and Slater, J.H. (eds), *Microbial Interactions and Communities*. Academic Press, London, pp. 13–44.

Bull, A.T., Huck, T.A. and Bushell, M.E. (1990) Optimization strategies in microbial process development and operation. In: Poole, R.K., Bazin, M.J. and Keevil, C.W. (eds), *Microbial Growth Dynamics*. IRL Press, Oxford, pp. 145–168.

De Weerd, K.A., Mandelco, L., Tanner, R.S., Woese, C.R. and Suflita, J.M. (1990) *Desulfomonile tiedjei* gen. nov. and sp. nov., a novel anaerobic, dehalogenating, sulfate-reducing bacterium. *Archives of Microbiology* 154, 23–30.

Gascoyne, D.J., Connor, J.A. and Bull, A.T. (1991) Isolation of bacteria producing siderophores under alkaline conditions. *Applied Microbiology and Biotechnology*, in press.

Giovannetti, L., Ventura, S., Bazzicalupo, M., Fani, R. and Materassi, R. (1990) DNA restriction fingerprint analysis of the soil bacterium *Azospirillum*. *Journal of General Microbiology* 136, 1161–1166.

Gochnauer, M.B., Leppard, G.G., Komeratat, P., Kates, M., Novitsky, T. and Kushner, D.J. (1975) Isolation and characterization of *Actinopolyspora halophila* gen. et sp. nov., an extremely halophilic actinomycete. *Canadian Journal of Microbiology* 21, 1500–1511.

Goodfellow, M. and O'Donnell, A.G. (1989) Search and discovery of industrially-significant actinomycetes. In: Baumberg, S., Hunter, I.S. and Rhodes, P.M. (eds), *Microbial Products: New Approaches*. Cambridge University Press, Cambridge, pp. 343–383.

Hawksworth, D.L. (1991) The fungal dimension of biodiversity: magnitude, significance, and conservation. *Mycological Research* 95, 441–452.

Hütter, R. (1982) Design of culture media capable of provoking wide gene expression. In: Bu'Lock, J.D., Nisbet, L.J. and Winstanly, D.J. (eds), *Bioactive Microbial Products: search and discovery*. Academic Press, London, pp. 37–50.

International Union for the Conservation of Nature and Natural Resources (1988) *Convention on the Conservation of Biological Diversity*. International Union for Conservation of Nature and Natural Resources, Gland.

International Union for the Conservation of Nature and Natural Resources (1980) *World Conservation Strategy*. International Union for Conservation of Nature and Natural Resources, Gland.

International Union for the Conservation of Nature and Natural Resources (1989) *From Strategy to Action. The IUCN response to the Report of the World Commission on Environment and Development*. International Union for Conservation of Nature and Natural Resources, Gland.

Jannasch, H.W. (1989) The microbial basis of life as deep-sea hydrothermal vents. *American Society for Microbiology News* 55, 413–416.

Lee, S.Y., Mabee, M.S. and Jangaard, N.O. (1978) *Pectinatus*, a new genus of the family Bacterioidaceae. *International Journal of Systematic Bacteriology* 28, 582–594.

McNeely, J.A., Miller, K.R., Reid, W.V., Mittermeier, R.A. and Werner, T.B. (1990) *Conserving the World's Biological Diversity*. International Union for Conservation of Nature and Natural Resources, Gland.

Mielenz, J.R., Jackson, L.E., O'Gara, F. and Shanmugan, K.T. (1979) Fingerprinting bacterial chromosomal DNA with restriction endonuclease EcoR1: comparison of *Rhizobium* spp. and identification of mutants. *Canadian Journal of Microbiology* 25, 803–807.

Netzer, W.J. (1990) Emerging tools for discovering drugs. *Bio/Technology* 8, 618–622.

Nisbet, L.J. and Porter, N. (1989) The impact of pharmacology and molecular biology on the exploitation of microbial products. In: Baumberg, S., Hunter, I.S. and Rhodes, P.M. (eds), *Microbial Products: New Approaches*. Cambridge University Press, Cambridge, pp. 309–342.

National Science Board (1989) *Loss of Biological Diversity: a global crisis requiring international solutions*. National Science Board, Washington, DC.

Omura, S. (1981) Screening of specific inhibitors of cell wall peptidoglycan synthesis. In: Ninet, L., Bost, P.E., Bouanchand, D.H. and Florent, J. (eds), *The Future of Antibiotherapy and Antibiotic Research*. Academic Press, New York, pp. 389–405.

Omura, S. (1986) Philosophy of new drug discovery. *Microbiological Reviews* 50, 259–279.

Pace, N.R., Stahl, D.A., Lane, D.J. and Olsen, G.J. (1986) The analysis of natural microbial populations by ribosomal RNA sequences. *Advances in Microbial Ecology* 9, 1–55.

Parkes, R.J. (1982) Methods for enriching, isolating and analysing microbial communities in laboratory systems. In: Bull, A.T. and Slater, J.H. (eds), *Microbial Interactions and Communities*. Academic Press, London, pp. 45–102.

Raven, P.H. (1988) Our diminishing forests. In: Wilson, E.O. (ed.), *Biodiversity*. National Academy Press, Washington, DC, pp. 119–122.

Rifai, M.A. (1989) Astounding fungal phenomena as manifestations of interactions between tropical plants and microorganisms. In: Lim, G. and Katsuya, K. (eds), *Interactions Between Plants & Microorganisms*. [Proceedings of a JSPS-NUS Inter-Faculty Seminar] National University of Singapore, pp. 1–8.

Rogall, T., Flohr, T. and Bottger, E.C. (1990) Differentiation of *Mycobacterium* species by direct sequencing of amplified DNA. *Journal of General Microbiology* 136, 1915–1920.

Schindler, P. (1980) Enzyme inhibitors of microbial origin. *Philosophical Transactions of the Royal Society, B* 290, 291–301.

Shelton, D.R. and Tiedje, J.M. (1984) Isolation and partial characterization of bacteria in an anaerobic consortium that mineralizes 3-chloro-benzoate acid. *Applied and Environmental Microbiology* 48, 840–848.

Sneath, P.H.A. (1979) BASIC program for identification of an unknown with presence-absence data against an identification matrix of percentage positive characters. *Computers and Geosciences* 5, 195–213.

Sneath, P.H.A. (1980) BASIC program for the most diagnostic properties of groups from an identification matrix of percent positive characters. *Computers and Geosciences* 6, 21–26.

Sorensen, B., Falk, E.S., Wisloff-Nilsen, E., Bjorvatn, B. and Kristiansen, B.E. (1985) Multivariate analysis of *Neisseria* DNA restriction endonuclease patterns. *Journal of General Microbiology* 131, 3099–3104.

Umezawa, H. (1972) *Enzyme Inhibitors of Microbial Origin*. University Park Press, Baltimore.

Verstraete, W. and Huysman, F. (1988) Environmental biotechnology. Future applications and needs. *Biotech-Forum* 5, 357–360.

Ward, D.M., Weller, R. and Bateson, M.M. (1990) 16S rRNA sequences reveal numerous uncultured microorganisms in a natural community. *Nature* 345, 63–65.

Wilson, E.O. (1988a) The current state of biological diversity. In: Wilson, E.O. (ed.), *Biodiversity*. National Academy Press, Washington DC, pp. 3–18.

Wilson, E.O. [ed.] (1988b) *Biodiversity*. National Academy Press, Washington DC.

Woese, C.R., Kandler, O. and Wheelis, M.L. (1990) Towards a natural system of organisms: proposal for the domains Archaea, Bacteria and Eucarya. *Proceedings of the National Academy of Sciences, USA* 87, 4576–4579.

Zehr, J.P. and McReynolds, L.A. (1989) Use of degenerate oligonucleotides for amplification of the nifH gene from the marine cyanobacterium *Trichodesmium thiebantii*. *Applied and Environmental Microbiology* 55, 2522–2526.

Discussion

Samways: Microbial researchers have an advantage over entomologists in that material can be collected in the field and then stored and revived when necessary. This is particularly important in tropical countries with high levels of endemism and where biotic materials are being lost.

Bull: Preservation of microorganisms is indeed an important aspect, but I am concerned over the genetic instability that can arise during maintenance in culture collections. There may nevertheless be occasions when it is desirable to go back to wild-type isolates to screen for particular attributes. Conservation in culture collections has to be seen as complementing the conservation of microorganisms in the wild.

Samways: Can entomologists and biocontrol workers look forward to using DNA technology to 'bring to life' or recover potential beneficial organisms rather as microbiologists are able to do today? DNA can already be used to 'fingerprint' dried insect specimens so in the distant future perhaps it could be used to resuscitate them.

Bull: I am sure that genetic analysis will become increasingly important in all groups of organisms.

Hawksworth: With respect to the maintenance of microorganisms in culture collections, many of these can be preserved without any risk of genetic change in liquid nitrogen. Using cryomicroscopy, protocols can be developed to tailor methods to particular strains. It must also be borne in mind that many microorganisms have *never* been grown in pure culture; only about 17% of the known species of fungi – and less than 1% of the estimated species – are represented in the world's culture collections (see p. 24). What microbiologists isolate is dependent on the procedures and media used. Isolates obtained are always only a subset of those really present. Even some potentially exploitable microorganisms, such as the myxobacteria, are currently difficult or impossible to grow, but techniques can often be developed to overcome such barriers when target organisms have been identified. You should however not be left with the impression that microbiologists have overcome all the problems.

Bull: I certainly endorse Professor Hawksworth's sentiments. The techniques available now, particularly immunofluorescence staining and DNA probes, mean that we are only now starting to be able to quantify the extent of the microbial world we do not know.

Eighteen

Genetic Engineering and Biodiversity

J.E. Beringer, P.K. Hayes & C.M. Lazarus, *Department of Botany, University of Bristol, Woodland Road, Bristol BS8 1UG, UK*

ABSTRACT: The impact of genetic engineering on biodiversity, and the use of recombinant DNA technology in assessing biodiversity are discussed. The transfer of genes between organizations by genetic engineering techniques will be used to impart new characteristics to crops. Cloned DNA sequences are also probes for desired genes in other organisms, and can be used to follow the fate of genes in crosses. The great number and diversity of microorganisms, and also their ability to perform potentially exploitable biochemical processes, means that they are of especial interest to the genetic engineer. However, a consequence of the ready availability of genes and the ability to synthesize DNA sequences is that the maintenance of biodiversity could be seen as less important. The introduction of genetically manipulated organisms into the environment is not without risk, and must be controlled by regulatory committees. Microorganisms are being manipulated to form molecules that are otherwise difficult to produce, and could therefore threaten traditional sources, including agricultural ones. While the potential for improvements in agricultural production from genetic engineering is enormous, the long-term survival of native species should not be prejudiced by a lack of concern for the need to maintain genetic diversity.

Introduction

In this contribution we concentrate on the impact of genetic engineering on biodiversity, and on the use of recombinant DNA technology in assessing biodiversity. Our emphasis on microorganisms and agriculturally important

The Biodiversity of Microorganisms and Invertebrates: Its Role in Sustainable Agriculture.
Edited by D.L. Hawksworth. © CAB International 1991.

crop plants is a reflection of the amount of work being carried out on these groups; many of the points made are equally applicable to invertebrates.

Genetic engineering

Genetic engineering allows the isolation, modification and reintroduction of genes into living organisms. Because natural gene transfer is not involved, the recipient need not be related to the source organism. Consequently, it is possible to introduce genes for attributes that are not normally found in a given species. Once isolated it is possible to make modifications to genes so that the timing, location, and amount of expression in any host is as desired by the experimenter. This ability is extremely important if, in future, we are to be able to produce efficient and safe organisms for commercial use in the environment.

Although it is theoretically possible to isolate and manipulate genes from any organism, a major limitation to the development of the technology at present is that relatively few species are amenable to existing methods for introducing DNA. However, the number of species for which transformation of DNA has been demonstrated increases almost monthly, and it will probably be possible in the future to introduce DNA into any organism. Indeed, even the cereals which have been very difficult to manipulate, are responding to new techniques. Rice was the first major cereal to yield to the technology (Tamiya *et al.*, 1988), and much excitement has been generated recently by the regeneration of fertile transgenic maize plants (Gordon-Kamm *et al.*, 1990).

A much more important limitation to the use of genetic engineering in agriculture is that the characteristics of greatest importance to farmers, such as yield, quality, and pest resistance are often multigenic, and the number and functions of the genes involved is poorly understood. Unless there is a reasonable understanding of what genes are doing, it is very difficult to isolate them. If many genes contribute towards a particular trait, it may not be possible to identify and isolate all the individual genes required. The isolation and manipulation of a single gene contributing to a polygenic trait may be of little or no value.

When the biochemistry and genetics of a characteristic are not well-known, there is little opportunity to use genetic engineering to make useful modifications in the metabolic functions responsible for that characteristic. Gene cloning and mutagenic procedures associated with it can be utilized to attempt to identify the genes responsible and develop a better understanding of how organisms function, but such studies are heavily dependent on high-quality biochemical and physiological research. Perhaps one of the major limitations to the exploitation of genetic engineering in agriculture is the poor state of knowledge of the biochemistry of potentially useful organisms, and the relative funding imbalance between gene cloning and biochemistry.

At present, genetic engineering is concentrating on genes that are already available. For plants these include genes for herbicide, virus, and insect resistance (Gasser and Fraley, 1989), together with sequences of DNA whose insertion and expression in a host is designed to reduce the activity of existing genes, such as those involved in the ripening of fruit (e.g. Smith *et al.*, 1988; Hamilton *et al.*, 1990). There is considerable interest in genes involved in the nutritional quality of seeds, and this work will undoubtedly lead to significant changes in the relative value of different types of grains. While the genetic manipulation of mammals and fishes is relatively well-established, there are few agronomically useful characteristics, with the possible exception of growth hormones, that are likely to be developed soon.

Another very important aspect of molecular genetics involves the use of cloned DNA sequences as probes. The basis for this technology is that when double-stranded molecules of DNA are heated, or exposed to alkaline conditions, the two strands separate, but will come together again under appropriate conditions. Nucleic acid molecules from the organisms of interest, often after digestion with a restriction endonuclease and electrophoretic size fractionation through an agarose gel, are rendered single-stranded and then immobilized on a membrane support. The membrane is incubated in a solution containing the probe DNA (also single-stranded) under conditions that promote binding of the probe to any complementary sequence present on the membrane; unbound probe DNA is then removed by washing the membrane. The position of the bound probe can be detected because it will have been made radioactive, or it will have been conjugated to some other form of label.

DNA probes can be used to follow the fate of genes during crosses, to study population structure by the analysis of polymorphisms or genetic fingerprints, and to allow the assessment of species diversity in field samples. The use of DNA probes in evaluating microbial diversity and in investigating the population structure of microbial communities is not highly developed. For diversity measurements two approaches are currently in vogue. The first uses group- or species-specific oligonucleotide probes, conjugated to a fluorescent dye, allowing the identification of single cells by fluorescence microscopy or flow cytometry (Amann *et al.*, 1990*a*, *b*). The second involves the selective recovery of 16S-type ribosomal RNA genes from bulk nucleic acid preparations. This method makes use of specific oligonucleotide primers, either for DNA synthesis from the RNA template (Ward *et al.*, 1990), or for gene amplification using the polymerase chain reaction (Giovannoni *et al.*, 1990). A measure of taxonomic diversity in the sampled population is obtained by analysing the cloned genes either through the use of group- and species-specific probes or by partial DNA sequencing. The analysis of microbial population structure using DNA probes has been reported recently (Young and Wexler, 1988; Hartmann, 1989).

Biodiversity and gene manipulation

One of the great advantages of microorganisms is that they are so numerous and diverse. The world contains an extremely large, and mostly unstudied range of microorganisms. There are, for example, about $10^8 g^{-1}$ of soil, representing perhaps thousands of species. It is possible that microorganisms capable of carrying out most commercially interesting reactions are present in the environment; the problem for biotechnologists is to identify and isolate them.

There are few processes in agriculture that are not affected by microorganisms, particularly in relation to nutrition and disease. We exploit the useful microorganisms by culturing them and using them as inoculants. Good examples include mycorrhizal fungi, nitrogen fixing bacteria (such as *Rhizobium*), biological control agents (such as the bacterium *Bacillus thuringiensis*; Waage, Chapter 13), and microbes used to ferment foods (such as the yeasts and bacteria used for cheese making). An aim of genetic engineering is to modify such organisms to improve their activity. The only example to date of such a microorganism approved for human food is a yeast that has been modified so that the enzyme maltase is produced continuously, rather than only after the yeast is exposed to maltose. Traditionally such a yeast would have been produced by selection of wild-type or mutated yeasts and then looking for natural variation. Presumably, as more genes become available for manipulation and techniques improve, there will be less interest in selecting natural variants.

An important consequence of the ready availability of genes, and, indeed, our ability to synthesize sequences of DNA and manufacture genes, will be that biodiversity will become less important to genetic engineers. If screening for natural variation is neglected there will be little need to maintain the biodiversity needed to provide useful organisms. It is unlikely that neglect will seriously affect microbial diversity and thus reduce the opportunity for future generations to isolate useful microorganisms.

The situation is rather different for invertebrates. Many of these are relatively large and are not widely dispersed. For example there is a real risk that freshwater crayfish in the UK may be displaced, and perhaps lost, because an introduced American species has been so successful (Stubbs, 1988). If genetically modified species of invertebrates are introduced in future that are more aggressive than native species, there is a risk that there will be serious implications for the maintenance of biodiversity.

Such problems may be made much worse if there is widespread use of genetically engineered plants carrying toxin genes to confer resistance to insect pests. Not only might such plants kill benign insects feeding on crops, but also the transfer of such genes to closely related weed species could lead to the widescale loss of insect species. Such a situation is, fortunately, unlikely because there is probably little chance that the inheritance of such genes would confer sufficient selective advantage to encourage their widespread dissemination in weedy species. However, it is going to be necessary for regulatory committees to take special care to ensure that problems of this type do not arise. Another problem that has been anticipated is that the use of transgenic crop plants relying on the expression of foreign proteins to protect them from pests may increase the risk of resistance arising among natural populations of the pest. A possible way around this problem is to use combinations of protecting factors to reduce drastically the likelihood of spontaneous resistance arising in the field. For example, the use of *B. thuringiensis* endotoxins (Vaeck *et al.*, 1987) could be complemented by co-expression of the cowpea trypsin inhibitor protein (Hilder *et al.*, 1987). Present interest in low input and sustainable agriculture will undoubtedly accentuate the need for biological solutions to problems that are presently solved by chemical means. It is essential that the hazards of natural and genetically modified biological control agents are assessed properly. Chemicals do at least have the advantage that they are not self-replicating and therefore mistakes can be contained.

Indirect risks to biodiversity

An important consequence of genetic manipulation of microorganisms is that it is becoming attractive to use them to manufacture molecules that are normally difficult to produce, or have to be imported from other countries. Sugar is an excellent example. For many years now cane producers have been threatened by sugar beet production within the European Community, and more recently by the development of technology that allows for the efficient enzymatic conversion of corn starch to high fructose syrup (Chaplin and Bucke, 1990). Within the European Community, quota limits on these sources of sugar protect exports from developing countries; but for how long? Genetic engineering also enables the isolation of genes for natural products until now exclusively produced in tropical countries. For example, vanilla and chocolate may at some time in the near future be produced by micro-organisms growing in fermenters or by temperate crops which have been genetically engineered to carry the relevant genes.

The threat to agriculture in developing countries is complex. First, there is the loss of markets for their exports. Second, is the dependence on restraints in production in temperate regions as occurs in sugar production, and the inevitable loss of independence that this causes. Third, is the change that will occur in agricultural practice as new crops are introduced and are farmed more intensively. This will lead to a loss in agricultural diversity and a further reduction in the widespread use of native crop species.

Genetic engineers have a strong vested interest in natural diversity because this provides the pool of genes needed for future exploitation. For example, there is considerable interest in a protein from the Brazil nut because it is high in sulphur amino acids. The gene has been isolated and is being studied as a potential way of improving the quality of seeds that are low in these amino acids.

Looking ahead 50 years, it is hard to predict how extensively genetically engineered organisms will be used in agriculture. It is likely that certain modes of disease resistance or improvements in nutritive value will be used widely. If the technology proves to be as successful as predicted, it may be applied to a few indigenous varieties in developing countries, but this will inevitably lead to a concentration on their use and a reduction in crop diversity. The extensive use of such 'improved' organisms will greatly increase the prospects for the rapid development and spread of diseases.

The potential for great improvements in agricultural production arising from the exploitation of genetic engineering is enormous. It is essential that the long-term survival of native species is not prejudiced by lack of concern for the need to maintain genetic diversity.

References

Amann, R.I., Krumholz, L. and Stahl, D.A. (1990*a*) Fluorescent-oligonucleotide probing of whole cells for determinative, phylogenetic, and environmental studies in microbiology. *Journal of Bacteriology* 172, 762–770.

Amann, R.I., Binder, B.J., Olsen, R.J., Chisholm, S.W., Devereux, R. and Stahl, D.A. (1990*b*) Combination of 16S rRNA-targeted oligonucleotide probes with flow cytometry for analyzing mixed microbial populations. *Applied and Environmental Microbiology* 56, 1919–1925.

Chaplin, M.F. and Bucke, C. (1990) *Enzyme Technology*. Cambridge University Press, Cambridge.

Gasser, C.G. and Fraley, R.T. (1989) Genetically engineering plants for crop improvement. *Science* 244, 1293–1299.

Giovannoni, S.J., Britschgi, T.B., Moyer, C.L. and Field, K.G. (1990) Genetic diversity in Sargasso Sea bacterioplankton. *Nature* 345, 60–63.

Gordon-Kamm, W.J., Spencer, T.M., Mangano, M.L., Adams, T.R., Daines, R.J., Start, W.G., O'Brien, J.V., Chambers, S.A., Adams Jr. W.R., Willetts, N.G., Price, T.B., Machey, C.J., Krueger, R.W., Kausch, A.P. and Lemaux, P.G. (1990) Transformation of maize cells and regeneration of fertile transgenic plants. *Plant Cell* 2, 603–618.

Hamilton, A.J., Lycett, G.W. and Grierson, D. (1990) Antisense gene that inhibits synthesis of the plant hormone ethylene in transgenic plants. *Nature* 346, 284–287.

Hilder, V.A., Gatehouse, A.M.R., Sheerman, S.E., Barker, R.F. and Boulter, D. (1987) A novel mechanism of insect resistance engineered into tobacco. *Nature* 330, 160–163.

Hartmann, A. (1989) *Characterisation du Genome de Rhizobium et Bradyrhizobium au Niveau Moleculaire et son Utilisation en Ecologie Microbienne: Diversité des Populations Naturelles; Potential de Transfert de Plasmides.* PhD thesis, Université de Bourgogne.

Smith, C.J.S., Watson, C.F., Ray, J., Bird, C.R., Morris, P.C., Schuch, W. and Grierson, D. (1988) Antisense RNA inhibition of polygalacturonase gene expression in transgenic tomatoes. *Nature* 334, 724–726.

Stubbs, D. (1988) *Towards an Introductions Policy.* Wildlife Link, London.

Tamiya, K., Arimoto, Y., Uchimiya, H. and Hinata, K. (1988) Transgenic rice plants after direct gene transfer into protoplasts. *Biotechnology* 6, 1072–1074.

Vaeck, M., Reynaerts, A., Höfte, H., Jansens, S., De Beuckeleer, M., Dean, C., Zabeau, M., Van Montagu, M. and Leemans, J. (1987) Transgenic plants protected from insect attack. *Nature* 328, 33–37.

Ward, D.M., Weller, R. and Bateson, M.M. (1990) 16S rRNA sequences reveal numerous uncultured microorganisms in a natural community. *Nature* 345, 63–65.

Young, J.P.W. and Wexler, M. (1988) Sym plasmid and chromosomal genotypes are correlated in field populations of *Rhizobium leguminosarum*. *Journal of General Microbiology* 134, 2731–2739.

Discussion

Haines: Many people are concerned about the release of genetically manipulated organisms into the environment. Who, nationally or internationally, is leading in the development of mechanisms and protocols to safeguard the environment?

Beringer: The European Economic Community (EEC) has directives, and the Organization for Economic Co-operation and Development (OECD) is developing guidelines for its member states. On the basis of past experience, many developing countries can be expected to follow the OECD guidelines. The United Nations has a working group on the release of genetically engineered organisms, which is charged with making recommendations for use in developing countries. It does concern me that most small countries are not going to be in a position to manage their own regulatory committees; the establishment of regional regulatory groups is therefore being considered to advise on the desirability of releases on a regional and not a country basis, as will also occur in Europe.

Waage: It is now a fact that resistance genes to *B. thuringiensis* develop in insects; this is known in Asia and the USA, and resistance is spreading due to the overuse of this organism as a pesticide. Constant exposure to pesticides or other agents is the quickest way to get resistance developed. While the philosophy of inserting resistance genes into crop plants might seem ideal, this is contrary to practices with agrochemicals and biocontrol agents that are used only where needed. As resistance is then more likely to develop in pest organisms, does this not caution against the inclusion of resistance genes in crops?

Beringer: Consideration is being given to this. Genes are only expressed when switched on, and we now have access to factors which switch on genes. The *B. thuringiensis* gene could be placed next to a regulatory sequence which is only switched on when the plant is attacked; this could even be linked to attacks on particular parts of plants, or when leaves are developing. As more becomes known about gene regulation, more can be done to target gene expression to a defined period of time. For many insect pests in many plants there is the prospect of a specific gene expression just at the optimal time for insect infestation of that part of the plant. The rest of the plant would not express the toxin and thereby greatly reduce the development of resistance.

Lynch: The first genetically engineered strain of *B. thuringiensis* with an elevated titre of the crystal toxin was patented by my Institute. However, we have been involved in an international strain selection programme which has led to the discovery of wild-type strains with even greater amounts of the toxin than the genetically engineered ones. A positive feature of genetic manipulation is that it can be an intermediary technology leading to new strategies for wild-type strain selection.

Somapale: The shortage of qualified microbiologists and scientists in related fields working in agriculture in developing countries such as Sri Lanka inhibits the study of relevant problems in this field. Could not the international scientific organizations emphasize the need for having them in their own countries as a priority, and also provide adequate assistance in training. Resources for training scientists in appropriate specific areas are currently limiting.

Nineteen

The Importance of Microbial Biodiversity to Biotechnology

L.J. Nisbet & F.M. Fox, *Xenova Limited, 545 Ipswich Road, Slough SL1 4EQ, UK*

ABSTRACT: Microbial biotechnology is the application of microbiological processes to make useful chemical products. With increasingly sophisticated methodologies available, it now has a wide range of applications. One such application, microbial natural product screening for new drugs, has been the backbone of the pharmaceutical and agrochemical industry for 60 years. The era of antibiotic discovery has given way to target-directed screens based on disease models in the search for low molecular weight molecules with receptor activity. This search for novel molecules is dependent on microbial biodiversity.

Microorganisms, fungi and bacteria, are more abundant than any other living organisms. They exhibit extraordinary diversity in terms of their nutrition, exploitation of ecological niches, interaction with other organisms and secondary metabolism. The types of molecules already identified from many microbes raise profound evolutionary and ecological questions; undoubtedly, some have evolved clear functions, for instance in chemical defence and cellular differentiation. An understanding of these functions advances the way forward for more sensitive, mode-of-action, screening programmes.

Introduction

Many prokaryotic and eukaryotic organisms, bacteria, cyanobacteria, fungi, plants, and lower animals produce metabolites with great potential for human use. The natural occurrence of these useful products is a feature of probably all living organisms, but often predominates in certain taxa. For

The Biodiversity of Microorganisms and Invertebrates: Its Role in Sustainable Agriculture.
Edited by D.L. Hawksworth. © CAB International 1991.

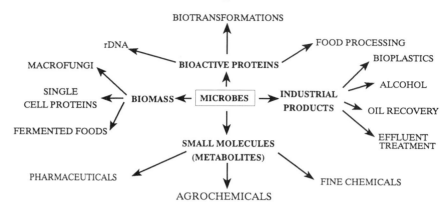

Figure 19.1. Biotechnological processes using microbes.

instance, among animals the most prolific producers are sponges, corals, and other marine invertebrates. They are extremely abundant in the plant kingdom but virtually unknown in the cyanobacteria. Several thousand, however, have been reported for bacteria and fungi (Berdy *et al.*, 1980).

The use of microorganisms in biotechnology is not a new science. Fermentation of alcohol and foods and the biotransformation of wine into vinegar have been practised since 5 000 BC, but recent developments in sophisticated equipment and biochemical methods has led to an escalation of research. Microbial biotechnology now has an enormous range of applications. It is used for the production of pharmaceuticals, fine chemicals and agrochemicals, for single cell proteins for animal feeds, for enzyme production, biopolymers, in effluent and waste treatment and in the genetic engineering of human proteins (Fig. 19.1).

The success of microbial biotechnology is dependent on the biodiversity of microorganisms and the array of chemical molecules produced during primary and secondary metabolism. This contribution discusses three aspects of biotechnology: the use of microorganisms in the discovery of chemical leads for the pharmaceutical and agrochemical industries, the rationale behind screening of microbial metabolites and the value of microbial biodiversity in the new mode-of-action approaches to drug discovery.

Drug discovery: Natural product screening

There are three ways by which novel molecules for pharmaceutical and agrochemical drugs can be discovered: synthetic chemistry, biopharmaceutical research, and the screening of naturally occurring molecules isolated from microorganisms, plants, and other natural sources. Synthetic chemistry is limited because it requires basic structures from the outset, and to date has not yielded a wide variety of novel structures because chemists need a starting point in the form of chemical templates. Recombinant DNA technology enables the production of molecules about which something is already known and has had recent success in producing modified antibiotics (Hopwood *et al.*, 1985). Microbial screening for natural products is a technology which really began in 1929 as a result of the discovery of penicillin, a secondary metabolite from *Penicillium chrysogenum*; indeed, its discovery along with cephalosporin from the fungus now known as *Acremonium chrysogenum* in 1945 laid the foundation of today's pharmaceutical companies, and established microbial screening programmes on a global scale. Since the 1940s, over 10 000 antibiotics have been discovered.

Microbial screening can be approached in two ways. Zahner *et al.* (1982) argue in favour of a non-target directed approach on the basis that only a fraction of natural microbial metabolites are known and screening techniques should be aimed at isolating new molecules irrespective of their activities. New structures have been successfully isolated in this manner, and include the nikkomycins which are chitin synthetase inhibitors inhibiting spore formation and which may be developed into new insecticides. On the other hand, Demain (1987) and others argue that although such non-directed screening will be successful in the isolation of chemically diverse metabolites, it will have a lower probability of finding commercial utility than will the candidates from target-directed screens.

The search for antibiotics during the last 40 years is giving way to an era of sophisticated assay technologies and opening up new therapeutic opportunities for microbial metabolites. Automation is now available to facilitate screening on a high throughput basis, and sensitive, specific and simple target-directed screens based on *ex vivo* models have been developed. Some of the new technologies which have facilitated screen design include the use of chromogenic substrates, radioligands, immunoassays such as ELISA, and microtitre systems which enable the automation of assay preparation, washing, harvesting and reading. The future holds prospects for the use of biosensors for enzyme inhibition, receptor ligand assays, and molecular sizing for the specific selection of small molecules.

Some of the major target-orientated screens include the search for antifungal, antitumour and antiviral agents; non-toxic anti-fungals are clearly needed to replace or supplement those presently in use such as amphotericin B and griseofulvin in human therapy, and kasugamycin and polyoxins in agriculture. Recently developed anti-fungal screens are investigating natural products that induce morphological abnormalities, and ecological, field-based research is a valuable way to identify organisms expressing antifungal activity. Co-ordinated research between pharmaceutical, agrochemical and biological research programmes may well advance the way forward.

Screening programmes designed to detect enzyme inhibitors have identified commercially important products such as the herbicide biolaphos, and mevinolin, a human cholesterol-reducing agent. Such enzyme assays from high throughput screening programmes should also lead to new antiparasitic products. Infectious diseases and famine remain the major impediments to progress in developing nations; in 1985 the synergistic action of malnutrition and infectious disease killed an estimated 6–7 million children under the age of 5 years in Africa. Tropical eukaryotic parasitic infections are due to blood and tissue protozoa, spirochaetes, and helminths with complex life-cycles and include schistosomiasis, filariasis, malaria, and trypanosomiasis; new vaccines and compounds are clearly needed. The concept that microbial secondary metabolites could be useful against parasites is supported by the potent activity of the semi-synthetic drug, ivermectin (Egerton *et al.*, 1979). Although originally considered solely for agricultural use against helminth infections in farm animals, ivermectin has now been shown to act as an effective agent against human onchocerciasis (river blindness), a parasitic disease affecting 20–40 million people.

Diversity of natural products in microbes: The rationale for microbial screening

Microorganisms produce chemicals as a result of primary and secondary metabolism. Primary metabolism involves catabolic and anabolic pathways to assemble biosynthetic precursors into the essential macromolecules of cellular structure and function and provide energy for chemical reactions. Life cannot exist without primary metabolism and the intermediates and products are essentially identical for all organisms. Secondary metabolism, on the other hand, involves metabolic pathways for the synthesis of compounds (natural products) that are not essential for the growth of the producing organism but they are derived solely from the precursors and energy generated through primary metabolism. The extraordinary feature of secondary metabolites lies in the fact that they arise from a very limited number of key intermediates (acetyl-coenzyme A, shikimate, amino acids and glucose) but the end result is a huge and diverse array of compounds. Over 3 000 antibiotic secondary metabolites have been recognized in the actinomycete genus *Streptomyces* alone, indeed the single species *Streptomyces griseus* can be induced to produce more than 50 antibiotics in laboratory culture.

The production of secondary metabolites is often genotypically and phenotypically specific (Bu'Lock, 1975) and this has been exploited in some plant and actinomycete taxa for taxonomic purposes. In artificial cultures, their production can be manipulated by alteration of the environmental and nutritional conditions to induce maximum expression. Given the enormous potential of secondary metabolite production in microorganisms, and the opportunities available for manipulation of the types and quantities produced in laboratory fermentation systems, the natural product scientist has a tremendous resource at hand for the discovery of new chemical leads for biotechnological application.

Much of our knowledge of the production and regulation of secondary metabolites comes from the study of commercially important organisms. The abundance of secondary metabolites and their diversity has led to speculation about their existence and their function, if any, to the producer organism. Zahner *et al.* (1983) proposed that secondary metabolism is a diffuse field where biochemical evolution is taking place continuously. The existence of this 'playground' is dependent mainly on the supply of surplus precursors and energy from primary metabolism, so that the evolution can proceed in all directions provided that the metabolites are not toxic to the producing organism. Such a biochemical playground requires great genetic flexibility and the patchy distribution of organisms exhibiting secondary metabolism has led to interesting questions. For instance, a glance at the occurrence of secondary metabolites in unrelated organisms separated very early in evolution, for example, β-lactams in both actinomycete bacteria and ascomycete fungi, has led to the speculation that the biosynthetic pathways leading to these products have been acquired by direct transfer of the corresponding genes. It is conceivable that this event might have taken place during a symbiont–host or parasite–host interaction, or simply by close contact and uptake of genetic material during prolonged growth in the same habitat such as soil.

The transfer of genetic materials is a well-established fact for the transfer of the Ti plasmid from *Agrobacterium tumefaciens* to its plant host (Chilton, 1980). If the assumption of genetic transfer were true for the β-lactam genes, one would expect identical biosynthetic pathways with identical or very similar enzymes, and, indeed, this is the case (Baldwin *et al.*, 1981). On the other hand, the occurrence of very similar metabolites among distant taxa could also result from the accumulation of the same intermediary precursors and this is thought to be likely for the presence of terpenoids to be found in basidiomycete fungi, plants, and marine animals.

According to this 'endosymbiont theory', it would seem plausible that the continuous evolution of biochemical pathways in secondary metabolism will lead to compounds which confer an advantage to the producing organism and the necessary biochemical information has, on occasion, been transferred to other organisms. This theory proposes that during evolution whole organisms, bacteria or cyanobacteria with their complete genetic make-up and metabolism, were incorporated into early eukaryotic cells and developed into mitochondria and chloroplasts. The rationale for this process was the usefulness of part of the prokaryotic metabolism to the host and the huge amount of time saved by making use of pre-existing genetic information instead of developing it *de novo*. During this process, genes for the enzymatic sequences were kept and information that conferred no advantage was eliminated.

Further, this suggests that once a useful biochemical pathway has been invented in secondary metabolism it can be utilized by other organisms. Certainly, studies of the ecological biochemistry of plant–animal–microbial interactions indicates that secondary metabolites have evolved a variety of clear functions in defence, cellular differentiation, and survival; indeed their functional role can perhaps only be clearly seen in relation to knowledge of how organisms interact and co-evolve. The recognition of function and co-evolution of secondary metabolites has important implications for pharmaceutical, agrochemical, and other biotechnological research, particularly with the increasing realization that co-evolution is not limited to plants and 'lower' organisms but can be extended to the 'higher' animal kingdom as well.

There are some interesting parallels between microbial secondary metabolites and certain differentiated animal cells which produce what are known as 'luxury molecules' because they are not essential to the viability of the cells producing them (Weinberg, 1983). Myosin and fibrinogen, for example, are not essential for muscle or liver cells; haemoglobin, insulin, and thyroxin, as well as their associated mRNAs, repressors and derepressors are other examples. Many luxury metabolites are proteins produced in high concentrations in specialized cells: for example, ovalbumin, globin, and fibroin. These specialized molecules and the mRNAs that encode them can be analysed by new biochemical methods and it has been surprising to discover there are intervening sequences in the genes for these luxury molecules. This finding is significant in the understanding of eukaryotic development. The differential activation and expression of genes can be under a variety of control mechanisms. Transcriptional control is the principal mechanism for genetic control in prokaryotes. It is now firmly established that it is also the principal mechanism for the control of genes for globin, ovalbumin, and fibroin. The exact mechanisms are unknown, but they may involve direct modification of the DNA.

The analogy between microbial and mammalian biochemical pathways has been extended by Roth *et al.* (1986) who propose that cell regulatory mechanisms in mammals have evolved from microbial sources. There are a number of fungal hormones for instance involved in sexual reproduction and some of these have a structural similarity to mammalian sex hormones. For example testosterone and oestradiol in humans are structurally similar to the steroidal sex hormones in the fungal genus *Achlya*. The two peptide mating factors in the yeast *Saccharomyces cerevisiae* may have hormone-like actions analogous to their mammalian counterparts. These and other hormonal molecules have features characteristic of receptor binding activity. Roth *et al.* (1986) have suggested that specialized vertebrate cells such as neurons, blood, tumour, and immune cells have evolved recently; the hormones, neuropeptides, and receptors by which they communicate, on the other hand, have more primitive and evolutionarily older microbial origins.

Evidence for the genetic transfer of information such as that seen between *A. tumefaciens* and its plant host is very scarce for microbial–mammalian interactions. One piece of information, however, provides an interesting link. In humans, chorionic gonadotrophin consists of two subunits coded by two genes. Some but not all bacteria that have been isolated from human tumours synthesize detectable amounts of the protein (Bachus and Affronti, 1981), but bacterial isolates from non-malignant tissues fail to produce the hormone. It is possible that the relevant genes could be present in the bacteria as a result of conservation or as an example of convergent evolution. However, as the genes appear to be expressed only in bacterial cells associated with tumour cells, the DNA may have been acquired from the host.

There is an increasing recognition that the pharmacological activity of microbial metabolites is by no means limited to antibiotic and toxigenic reactions. Demain (1981) has listed over 60 pharmacological activities from microbial secondary metabolites. The list is inevitably anthropocentric: scientists concentrate on compounds that will have clinical, industrial or agricultural application. Nevertheless, the inference is clear: microbial secondary metabolites display a wider range of physiological activity than has hitherto been suspected, and examination of the producing organisms in their natural habitats reveals some profound physiological and evolutionary questions.

There is a basic tenet in biology which seems to be frequently overlooked in discussions on the role of secondary metabolites 'in biology, nothing makes sense except in the light of evolution' (Dobzhansky, 1970). The diversity of chemical structures and biological activities displayed by microbial metabolites suggests that a diversity of ecological functions are plausible, selective pressures are operative in the natural environment, and co-evolutionary mechanisms may be highly significant.

Microbial biodiversity: Its value in drug discovery

Biodiversity is an ecological term used to describe the different macro- and micro-ecosystems within a geographical area, the diversity of different species within an ecosystem, and the genetic diversity within species. Within the context of this contribution, it is also regarded as the phenotypic and genotypic specificity of metabolite production. Intraspecific differences in the production of specific metabolites are often significant; they can certainly cause problems for the natural product scientist, but also greatly increase the potential for the discovery of novel chemical structures. There are several ways by which a directed organism isolation programme can be conducted in order to look for molecules of potential interest. Organisms can be system-atically isolated and screened according to their taxonomy, and although this may have some value for the actinomycetes, the higher taxonomic groupings of the fungi are generally artificial and bear little relation to the ecology and physiology of the groups. Secondary metabolite production in relation to substrate specificity and cytodifferentiation can also be studied and this can be field or laboratory based.

To date, natural product research has tended to concentrate on the isolation of molecules from actinomycetes, and this is particularly the case with the discovery of novel antibiotics in Japan. The actinomycetes are Gram-positive prokaryotes comprising 63 genera and are relatively well-documented because of their value in drug discovery. They are generally regarded as sporulating, soilborne organisms, but show considerable ecologi-cal and biochemical diversity. A number of key genera are adapted to symbiotic commensalism and extreme environments, and these groups have received little attention because they tend to be difficult to isolate and culture. Novel genera are still being discovered; an example is *Kibdelosporangium aridum* found in a bat cave in Arizona, which produces a glycopeptide antibiotic now developed for animal health (Jeffs and Nisbet, 1988). Some are associated with other organisms: *Frankia* is a genus of nitrogen fixing sym-bionts of alder and other woody dicotyledons, the genus *Actinomadura* is a human pathogen, and *Nocardia* 'rhodnii' is a cellulolytic actinomycete that inhabits the gut of the insect *Rhodnius prolixus* (Williams *et al.*, 1989).

Biodiversity, ecological niche exploitation, association with other organ-isms, and sheer abundance reaches its greatest complexity with the fungi though, unlike the actinomycetes, they have scarcely been explored for bioactivity. The fungal kingdom is composed of groups that have evolved along separate evolutionary paths, and they are an extraordinarily diverse group. There are about 69 000 species of fungi known, but well over 1 000 new species are being discovered each year, and a conservative world total may be in the region of 1.5 million (Hawksworth, 1991). A considerable proportion of the new records are from the tropics, and the lack of available data on the biodiversity of fungi in such complex and unexplored ecosystems to date is primarily due to the lack of mycologists working in the area. During an ecological study of tropical forest in the El Verde forest reserve in Puerto Rico, Lodge (1988) estimated that 20–30% of the agarics (i.e. the large fruiting basidiomycetes) she encountered are new species. Dennis (1970) indicated that in his fungal flora of Venezuela, which lists 2 412 species of mainly macrofungi, this is a 20-fold underestimate of the probable species total for that region.

Of the main fungal groups (Fig. 19.2), the Deuteromycotina are prob-ably the most thoroughly investigated for bioactivity, and this may be due at least in part to their relative ease of isolation from the field and their vigour in fermentation systems. These generally rapidly sporulating

THE FUNGAL KINGDOM

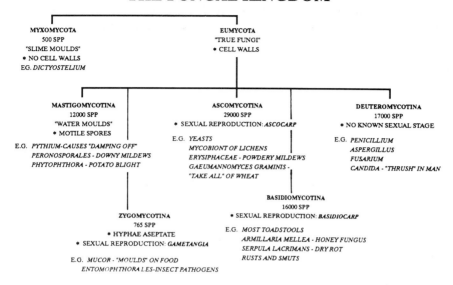

Figure 19.2. The fungal kingdom.

fungi lend themselves readily to current methods of batch culture, cryopreservation, and high-throughput screening. *Penicillium chrysogenum*, from which penicillin was first isolated, and *Aspergillus terreus*, the source of mevinolin (lovastatin), are examples of soil-inhabiting saprobic members of this group.

Several compounds with distinct pharmacological activity have already been established from basidiomycetes (Fig. 19.2): the antitumour properties of Shiitake (*Lentinus edodes*), the 'elixir of life' cultivated and marketed in the Far East is one of the best known. Indeed, extracts from fruit-bodies of basidiomycetes have been used for centuries by humans in food prepration, for medicinal purposes, as toxins, and as hallucinogens. However, unless the fruit-bodies can be produced in the laboratory they have only limited value for pharmacological research because field collection of large amounts of material is environmentally unsound and the results are difficult to reproduce. Recently, Anke *et al.* (1980), Anke and Steglich (1981), workers at Xenova Ltd (unpublished), and others, have begun to investigate the pharmacological and agrochemical bioactivity of secondary metabolites from the vegetative mycelium of basidiomycetes, and several novel molecules have been identified. They are a challenging group but the complexity of cytodifferentiation and the uniqueness of the dikaryon suggests that they have enormous potential in drug discovery.

A characteristic feature of the most prolific producers of secondary metabolites amongst microorganisms is their adaptability to changing environmental conditions such as substrate, pH, temperature, nutrient, oxygen, and water availability (Zahner *et al.*, 1983). The ability to use a wide variety of substrates of different composition which enter primary metabolism at different points leads to an enlargement of the available pool for secondary metabolism. On the other hand, microbes confined to extreme environments such as hot sulphur springs (*Sulfolobus* species), burning coal piles (*Thermoplasma* species), and those able to use substrates which cannot be metabolized by others (melanotrophic bacteria), do not seem to have an important secondary metabolism (Zahner *et al.*, 1983). Evolution has led them to perfectly adapt to their environment which is concomitant with a loss of adaptability. If this is true, isolation of organisms from habitats which are rapidly changing or where competition between organisms is intense may be productive sources of novel molecules. Directed organism isolation programmes can be based on such ecological considerations as habitat and nutritional requirements.

There is a substantial body of information available which links secondary metabolites with cytodifferentiation (Zahner *et al.*, 1983; Weinberg, 1983), and this can be an alternative approach to directed isolation. In fungi, as well as in actinomycetes, secondary metabolite production often appears to coincide with cellular differentiation: for example, production of cephalosporin and arthrospore formation in *Acremonium chrysogenum* (Martin and Demain, 1978), and ergot alkaloid production and formation of chlamydospores in *Claviceps purpurea* (Spalla and Marnati, 1978). In *C. purpurea*, alkaloid production is restricted to the sexual fruiting stages and the appearance of asexual conidia is associated with the loss of ergot-producing ability. Sclerotia of many fungi frequently contain metabolites that confer protection against predation in the natural environment; ergot alkaloids in *C. purpurea* appear to have this function. However, because fungal sclerotia do not form in the conventional fermentation systems of liquid shaken culture, many novel metabolites undoubtedly remain undetected.

Of all the fungi, the Basidiomycotina probably exhibit the most extensive differentiation, especially during the formation of the sexual, homokaryotic fruiting body from the heterokaryotic vegetative mycelium; the process involves a complex sequence of tissue differentiation, gene switching, and biochemical regulation. De Vries and Wessels (1984) have demonstrated differences in the types of molecules that can be detected in vegetative mycelium compared with that which has initiated sexual primordia. For instance, a range of low molecular weight polypeptides are produced in *Schizophyllum commune* during sexual fruit-body formation which are not produced in the vegetative state.

Alternatively, organism isolation programmes for natural product screening can look to the role of secondary metabolites in the natural environment based on the evidence that secondary metabolites have evolved a role in chemical defence for many organisms. Study of this phenomena can specificially direct the search for compounds that will be of pharmacological or agrochemical interest. An example of such directed searching is reported by Waterman (1990): *Myrica gale* is a shrub notable for the relative absence of predation by herbivorous insects and fungal pathogens – analysis of the volatile oils and phenolic compounds in the leaves has led to the isolation of a novel flavonoid with antifungal activity against *Penicillium citrinum*.

'Functional' secondary metabolites, that is compounds that provide advantages to the producer organism, reach their most complex, and often most bizarre, in fungal–plant and fungal–insect interactions. Shoot endophytes have recently attracted attention because of the economic losses associated with ryegrass staggers, fescue toxicosis, and related diseases of grazing ruminants (Clay *et al.*, 1985). The ryegrass endophyte *Acremonium loliae* asymptomatically infects ryegrass and produces a tremorgen toxin, lolitrem B, which causes neurological disorders in cattle and sheep. It also produces two other secondary metabolites: a plant growth hormone analogue, tryptophol, which increases plant biomass and increases drought resistance, and peramine, a natural insect antifeedant. Other fungal endophytes, for example, *Elytroderma torres-juanii* occurring in pine needles of *Pinus brutia* and *Rhabdocline parkeri* in Douglas fir needles, confer pronounced and well-proven herbivorous resistance on their hosts (Clay, 1988).

Other mycotrophic interactions provide fascinating evolutionary questions. Some smuts are seed-replacing, systemic fungal infections of grasses, and cause antagonistic biotrophic infections of major cereal crops. Characteristically, the smut infection affects only a portion of the seed crop and the fungus is 'confined' within the seed. As a dispersal mechanism, it seems probable that it has co-evolved with herbivores by producing trimethylamine present in the spores which is both a mammalian sex attractant and a spore germination inhibitor (Harborne, 1982). Mycorrhizal fungi may have evolved similar dispersal mechanisms; for example, the relationship between mycophagous rodents such as squirrels and hypogeous fungi such as truffles suggests a co-evolved mutualism in which for exchange of spore dispersal, these fungi offer 'ripe' fruit advertised by scent (Pirozynski and Malloch, 1988). The attractant appears to be a mammalian sex pheromone (Claus *et al.*, 1981), a potent incentive for rodents. Further, the evolution of seed burial by rodents may originate in the advantages of burying the seed cache near ectomycorrhizal roots in soil, the rhizosphere of which exerts an antibiotic action leading to 'safe' storage of the food supply. For the natural product scientist these questions of co-evolution, biodiversity, and niche adaptation are not merely interesting issues but provide important leads for the directed isolation of specialized groups of organisms that may be induced to produce pharmacologically active small molecules that have mammalian receptor activity. Ecological research such as that described above gives further support to the theories put forward by Roth *et al.* (1986) that vertebrates and microbes have close evolutionary links.

Fungi that parasitize insects are a taxonomically diverse group unique in their capacity to produce extracellular enzymes which enable penetration of the intact insect cuticle (Evans, 1987), and it is thought that they have evolved with their insect hosts over a considerable period. Once they have penetrated the host, the fungal pathogens are able to overcome host defence mechanisms, and the insect remains alive while the fungus multiplies within the host haemocoel. During the infection process some insect-parasitizing fungi produce neurotoxins which alter the behaviour of the host to the advantage of the fungus. For example, in the 'summit disease' *Melanoplus bivittatus* invades grasshoppers causing them to congregate at the tops of grass stems before dying thus facilitating the dispersal of fungal spores. This activity of diseased insects is abnormal; exposure increases the chance of predation by birds and it is suggested that the host is specifically programmed by the fungus which either interferes with the host nervous system directly or releases substances which induce behavioural changes indirectly (Evans, 1987). There are many such examples of fungal mediated behavioural changes in which the dispersal strategy of the fungus demands that the host continues its normal behavioural pattern thus ensuring that healthy individuals come into contact with the fungal inoculum. This is strikingly illustrated in *Massospora* infected male cicadas which still maintain their mating routine despite the loss of sexual organs (Evans, 1987). Fungal metabolites have seldom been studied in insects, but it is conceivable that there may be fungal products that are potent against insects but have a low toxicity to mammals. Potential examples are the destruxins of *Metarrhizium anisopliae* and beauvericin from *Beauveria bassiana*, which are toxic to insects but have not been implicated in mammalian toxicoses. Insect parasitizing fungi have also attracted considerable recent interest as potential sources of biological control agents.

Lichenization, the production of stable mutalistic symbioses between fungi and algae and/or cyanobacteria, is more prevalent than commonly realized, and provides an extraordinary example of biodiversity. Ascomycotina are the main mycobionts, and it is estimated that 13 500 fungal species may enter into lichen symbioses, many of which are obligate (Hawksworth, 1988). They are prolific producers of secondary metabolites, and because they often occur in extreme habitats such as boreal zones, tropical forest canopies and high altitudes, they provide an argument against Zahner *et al.*'s (1983) theory that secondary metabolite production is not associated with adaptation to extreme environmental niches. About 350 secondary metabolites have been characterized from lichens, and some of the products have been used by humans for centuries. In the 1950s many were screened in natural product research programmes, and some were shown to contain usnic acids which have antibiotic properties (Hawksworth and Hill, 1984)

Table 19.1. The numbers of species of microorganisms compared with those maintained in service culture collections (based on data compiled by D.L. Hawksworth).

Group	Number of species		Species in culture collections	
	Described	Estimate of total	Number	Percent of total
Algae	40 000	60 000	1 600	2.5
Bacteria	3 000	30 000	2 300	7.0
Fungi	69 000	1 500 000	11 500	0.8
Viruses	5 000	130 000	2 200	2.0

and other compounds with antitumour and antiviral properties. Some are still used in commercial preparations. The Greeks used lichens as a source of dyes and 'oak moss' and 'tree moss' are still used extensively in the manufacture of perfumes. These processes, however, use intact lichens and thus raise important conservational issues. A project funded by the Department of Trade and Industry was established in 1990 between four institutes in the UK: the University of Nottingham, the International Mycological Institute, Biocatalysts Ltd, and Xenova Ltd, to conduct research into the discovery of molecules from the mycobionts of lichens cultured in the laboratory. Only a small amount of lichen tissue is required for the initial isolation of the mycobiont, and thus avoids the destruction of the lichen in its natural habitat.

Mycorrhizal symbioses are to the Basidiomycotina what lichenization is to the Ascomycotina, and mycorrhizas may underlie the success of the basidiomycetes. Miller and Watling (1987) estimate that 3 500–5 000 taxa are involved. Like lichens, mycorrhizas show a remarkable capacity for niche exploitation and adaptation. They are important commercially in forest regeneration programmes, reclamation of toxic sites, in horticulture and agriculture especially in the tropics, and in commercial orchid growing. Ectomycorrhizas produce fruit-bodies that are characteristic of temperate woodland flora in autumn and include *Amanita* species, some fruit-bodies of which contain toxins and hallucinogens, and *Boletus* species, many of which are edible. Mycorrhizal roots exhibit antibiosis and may protect the host tree from pathogen attack. Bioactivity of the heterokaryotic mycelium has scarcely been studied because many require, as yet unknown, growth factors for successful laboratory culture.

Diversity of organisms and the molecular biological techniques now available provide an enormous genetic resource at our disposal, but effective conservation and preservation of this gene bank requires substantial culture collections. It is estimated there are 100 000 described species of microbes, but there are probably 10 times more in reality and less than 5% of these are currently held in collections (Table 19.1). The co-ordination and quality of such institutions requires skills in taxonomy, preservation techniques, and compilation and co-ordination of data, training in which is often disregarded in education. There is much information published on the loss of valuable microbial and botanical resources with the destruction of tropical and temperate habitats, but preservation of habitats is not sufficient in itself to ensure the protection of this resource for natural product research: the value of the gene banks themselves must be increasingly appreciated and utilized.

Conclusions

Despite its serendipitous nature, natural product screening has had a profound effect on the pharmaceutical and agrochemical industries. The era of antibiotic discovery based on random screening of secondary metabolites from microbes is being replaced by sophisticated, target-directed, mode-of-action screens. This new approach relies on the technological tools of modern molecular biology and involves the search for low molecular-weight molecules with specific activities. Concomitant with this new approach, knowledge of ecology and an appreciation of biodiversity is used to direct microbial isolation programmes which in turn can influence screen technology. Understanding the role of secondary metabolism in nature provides valuable leads. It is becoming increasingly realized that evolution links groups of organisms in a way that has hitherto been scarcely recognized. Natural product screening has traditionally been regarded as a 'needle-in-a-haystack' exercise, but recent research suggests that the search among microbes for molecules with receptor activity in mammalian systems has a previously unrecognized value and rationale behind it.

The success of such programmes relies on the biodiversity of microorganisms and the conservation of the genetic resource they provide. Global coordination of culture collections is needed. Habitats need to be preserved and an international code of conduct is required to bring about full co-operation and support between developing and developed nations so that the disciplines of ecology, taxonomy, and conservation can work alongside industry and the fields of molecular biology and biomedical science.

References

Anke, T.M., Kupka, J., Schramm, G. and Steglich, W. (1980) Antibiotics from basidiomycetes. X, Scorodonin, a new antibacterial and antifungal metabolite from *Marasmius scorodonius. Journal of Antibiotics* 83, 463–467.
Anke, T. and Steglich, W. (1981) Screening of basidiomycetes for the production of new antibiotics. In: Moo-Young, M., Robinson, C.W. and Vezina, C. (eds), *Advances in Biotechnology*. Pergamon Press, Toronto, vol. 1, pp. 35–40.
Bachus, B.T. and Affronti, L.F. (1981) Tumour-associated bacteria capable of producing a human choriogonadotropin-like protein in bacteria isolated from cancer patients. *Cancer* 41, 1217–1229.

Baldwin, J.E., Keeping, J.W., Singh, P.D. and Vallejo, C.A. (1981) Cell-free conversion of isopenicillin N into deacetoxycephalosporin C by *Cephalosporium acremonium* Mutant M-0198. *Biochemical Journal* 194, 649–651.

Berdy, J., Aszalos, A., Bostian, M. and McNitt, K.L. (eds) (1980) *Handbook of Antibiotic Compounds.* Vol. 2. *Macrocyclic Lactone (Lactam) Antibiotics.* CRC Press, Boca Raton.

Bu'Lock, J.D. (1975) Secondary metabolism in fungi and its relationship to growth and development. In: Smith, J.E. and Berry, D.R. (eds), *The Filamentous Fungi, Industrial Mycology.* Vol. 1. Edward Arnold, London, pp. 35–58.

Chilton, M.D. (1980) *Agrobacterium* Ti plasmids as a tool for genetic engineering in plants. In: Rains, D.W., Valentine, R.C. and Hollaender, A. (eds), *Genetic Engineering of Osmoregulation.* Plenum Press, New York, pp. 22–31.

Claus, R., Hoppen, H.O. and Karg, H. (1981) The secret of truffles: a steroidal pheromone. *Experentia* 37, 1178–1179.

Clay, K. (1988) Fungal endophytes of grasses: a defensive mutualism between plants and fungi. *Ecology* 69, 10–16.

Clay, K. Hardy, T.N. and Hammond, A.B. (1985) Fungal endophytes of grasses and their effect on an insect herbivore. *Oecologia* 66, 1–5.

de Vries, O.M.H. and Wessels, J.G.H. (1984) Patterns of polypeptide synthesis in non-fruiting monokaryons and a fruiting dikaryon of *Schizophyllum commune. Journal of General Microbiology* 130, 145–154.

Demain, A.L. (1981) Industrial microbiology. *Science* 214, 987–995.

Demain, A. L. (1987) New applications of microbial products. *Science* 219, 709–714.

Dennis, R.W.G. (1970) Fungal flora of Venezuela and adjacent countries. *Kew Bulletin, additional series* 3, i–xxxiv, 1–531.

Dobzhansky, T. (1970) *Genetics of the Evolutionary Process.* Columbia University Press, New York.

Egerton, J.R., Ostlind, D.A., Blair, L.S., Eary, C.A., Sohayda, D., Cifelli, S., Riek, R.F. and Campbell, W.C. (1979) Avermectins, a new family of potent anthelmintic agents: efficiency of the B1a component. *Antimicrobial Agents and Chemotherapy* 15, 372–378.

Evans, H.C. (1987) Mycopathogens of insects of epigeal and aerial habitats. In: Wilding, N., Collins, N.M., Hammond, P.M. and Webber, J.F. (eds), *Insect-Fungus Interactions.* Academic Press, London, pp. 205–238.

Harborne, J.B. (1982) *Introduction to Ecological Biochemistry.* 2nd edn. Academic Press, New York.

Hawksworth, D.L. (1988) Coevolution of fungi with algae and cyanobacteria in lichen symbioses. In: Pirozynski, K.A. and Hawksworth, D.L. (eds), *Coevolution of Fungi with Plants and Animals.* Academic Press, London, pp. 125–149.

Hawksworth, D.L. (1991) The fungal dimension of biodiversity: magnitude, significance, and conservation. *Mycological Research* 95, 441–452.

Hawksworth, D.L. and Hill, D.J. (1984) *The Lichen Forming Fungi.* Blackie, Glasgow.

Hopwood, D.A., Malpartida, F., Kieser, H.M., Ikeda, H., Duncan, J., Fujii, I., Rudd, B.A.M., Floss, H.G. and Omura, S. (1985) Production of 'hybrid' antibiotics by genetic engineering. *Nature* 314, 642–644.

Jeffs, P.W. and Nisbet, L.J. (1988) Glycopeptide antibiotics: a comprehensive approach to discovery, isolation, and structure determination. In: Daneo-Moore, L., Higgins, M.L., Salton, M.R.J. and Shockman, E.D. (eds), *Antibiotic Inhibition of Bacterial Cell Surface Assembly and Function*. American Society for Microbiology, Washington, DC, pp. 509–530.

Lodge, J. (1988) Three new *Mycena* species (Basidiomycotina, Tricholomataceae) from Puerto Rico. *Transactions of the British Mycological Society* 91, 109–116.

Martin, J.F. and Demain, A.L. (1978) Fungal development and metabolite formation. In: Smith, J.E. and Berry, D.R. (eds), *The Filamentous Fungi, Developmental Biology*. Vol. 3. Edward Arnold, London, pp. 426–450.

Miller, O.K. and Watling, R. (1987) Whence cometh the agarics? A reappraisal. In: Rayner, A.D.M., Brasier, C.M. and Moore, D. (eds), *Evolutionary Biology of the Fungi*. Cambridge University Press, Cambridge, pp. 435–449.

Pirozynski, K.A. and Malloch, D.W. (1988) Seeds, spores and stomachs: coevolution in seed dispersal mutualisms. In: Pirozynski, K.A. and Hawksworth, D.L. (eds), *Coevolution of Fungi with Plants and Animals*. Academic Press, London, pp. 227–247.

Roth, J., Leroith, D., Collier, E.S., Watkinson, A. and Lesnisk, M.A. (1986) The evolutionary origins of intercellular communication and the Maginot Lines of the mind. *Annals of the New York Academy of Sciences* 436, 1–11.

Spalla, C. and Marnati, M.P. (1978). Genetic aspects of the formation of ergot alkaloids. In: Hutter, R., Leisinger, T., Nuesch, J. and Wehrli, W. (eds), *Antibiotics and other Secondary Metabolites. Biosynthesis and Production*. Academic Press, London, pp. 219–232.

Waterman, P.G. (1990) Searching for bioactive compounds: various strategies. *Journal of Natural Products* 53, 13–22.

Weinberg, E.D. (1983) Comparative aspects of secondary metabolism in cell cultures of green plants, animals and microorganisms. In: Bennett, J.W. and Ciegler, A. (eds), *Secondary Metabolism and Differentiation in Fungi*. Marcel Dekker, New York, pp. 73–94.

Williams, S.T., Sharpe, M.E. and Holt, J.G. [eds] (1989) *Bergey's Manual of Systematic Bacteriology*. Vol. 4. Williams and Wilkins, Baltimore.

Zahner, H., Anke, H. and Anke, T. (1983) Evolution and secondary pathways. In: Bennett, J. W. and Ciegler, A. (eds), *Secondary Metabolism and Differentiation in Fungi*. Marcel Dekker, New York, pp. 153–175.

Zahner, H., Drautz, H. and Weber, W. (1982) Novel approaches to metabolite screening. In: Bu'Lock, J.D., Nisbet, L.J. and Winstanley, D.J. (eds), *Bioactive Microbial Products: search and discovery*. Academic Press, London, pp. 51–70.

Discussion

Robinson: Professor Nisbet has drawn attention to the potential of *Cordyceps* species as biocontrol agents. There is an extensive Chinese literature on *Cordyceps* as a medicinal herb, and it is widely used and available as such in South-East Asia. It is reported to be efficacious in the case of problems of the older male. In addition to its potential as a biocontrol agent, its pharmacological ingredients merit investigation in the West.

Twenty

The Universal Issue: Information Transfer

P.R. Scott, *CAB International, Wallingford, Oxfordshire OX10 8DE, UK*

ABSTRACT: Some aspects of information handling relevant to biodiversity and sustainable agriculture are reviewed, with an emphasis on the contribution of information technology. Factual information is considered first. The handling of extensive bibliographic data is discussed with reference to the use of electronic databases and CD-ROM technology. The need for consistent names or codes for organisms is reviewed. Culture collections are considered as a source of open-ended information about organisms. The handling of information on nucleotide sequences is discussed as a method of specifying the organism. Interpretive information is considered next, with emphasis on taxonomic database systems, and with reference to distribution maps and multimedia databases. Decision-making information is exemplified by a consideration of expert systems. Predictive information is discussed with particular reference to modelling the distribution of organisms in relation to their climatic preferences. For effective communication of information in relation to biodiversity, standardization of methodology and terminology is considered especially important.

Introduction

The purpose of this book, and the Workshop on which it is based, is the transfer of information: information to impart knowledge, to stimulate ideas, to provoke discussion, to develop new perspectives, to provide a basis for future action, and to give power to manage and control. The aim of this contribution is not to present more information about biodiversity and

The Biodiversity of Microorganisms and Invertebrates: Its Role in Sustainable Agriculture.
Edited by D.L. Hawksworth. © CAB International 1991.

sustainable agriculture, but to analyse some of the processes of information handling that are relevant to this field.

In the study of biodiversity, biotechnology and information technology are two of the arenas of most rapid technical advance. Examples discussed below illustrate some parallels between them. Biotechnology provides us with a new degree of control over certain aspects of the biological environment. Information technology is revolutionizing our ability to handle knowledge of that environment.

Four kinds of information are considered by reference to examples: (i) factual information; (ii) interpretive information; (iii) decision-making information; and (iv) predictive information.

Factual information

The theme of this book is the biodiversity of microorganisms and invertebrates that provides the ecological foundation for sustainable agriculture. Cataloguing and indexing that biodiversity is perhaps the most important single contribution that can be made in understanding its significance and in assessing the impact of environmental change on the diversity of organisms that play a role in land-use.

Bibliographic information

The most readily available source for most factual information about biodiversity is the primary literature of the life sciences. In compiling the CAB ABSTRACTS bibliographic database of the research and development literature of agriculture, forestry, and allied disciplines, CAB International regularly scans some 15 000 serial publications, together with some 5 000 books per annum. Source documents come from some 120 countries and are written in more than 50 languages. Although English is increasingly the international language of science, some 40% of the literature is in other languages. The CAB ABSTRACTS database, containing abstracts of the source literature with bibliographic details and index terms, is growing at the rate of some 150 000 records per annum. BIOSIS® providing a comparable service for a range of biological and medical sciences, scans some 9 000 serials, together with other publications, and its BIOSIS PREVIEWS database is growing by some 580 000 bibliographic records per annum.

These example statistics show the large scale of the task of maintaining bibliographic records in the field of agriculture and the life sciences, a substantial proportion of which have a bearing on biodiversity and its relationship to sustainable agriculture. Information technology is an essential management tool in both compilation and retrieval of this information.

Dextre Clarke (1988) analysed the uses and future of bibliographic database systems (in the context of a symposium on biosystematics), anticipating a trend towards diversity of media and presentation, and especially the probable growth of the use of CD-ROM technology. The spread of CD-ROM technology is currently entering a phase of rapid increase. Scott and Harris (1989) presented an analysis of responses to an international questionnaire about the use of crop protection information, which showed a great preponderance of usage of information on paper rather than in electronic form. While the printed page remains the most popular medium, the growth of CD-ROM usage will quickly render the details of that analysis out of date.

For a field as broad-ranging and lacking in precise definition as biodiversity, providing access to relevant information in bibliographic records makes substantial demands of indexing systems. An essential element in efficient indexing and information retrieval is the consistent use of terms, especially the names of organisms.

Names of organisms

Names have particular importance in information handling. The name of an organism provides the point of access to information about it. The name of a pest organism, for example, can be the key to information about its morphology, its genetics, its distribution, its ecology and epidemiology, its hosts, its natural enemies, its agricultural impact, and the means to control it.

The 1989 International Crop Protection Information Workshop (ICPIW) recommended that 'the production of thesauri and lists of recommended names for organisms of agricultural importance by . . . international organizations concerned with biological nomenclature should be encouraged in the interests of stability and ease of communication' (McDonald, 1989). Publications such as the *Index Kewensis, Zoological Record,* and *Index of Fungi* perform an essential service in maintaining records of published names of organisms. Examples of thesauri of recommended scientific names include the *CAB Thesaurus* (CAB International, 1990), which also covers all other descriptors used in indexing the CAB ABSTRACTS database, and a thesaurus of names of crops and crop pests based on the terms used to index the PESTDOC database of pesticide and related information (Derwent Publications, 1990). There is extensive scope for more detailed lists of preferred names of particular sets of organisms. The international initiative to prepare a 'list of names in current use' for the approximately 36 500 genera covered by the International Code of Botanical Nomenclature, with the intention to extend the list to specific names later, is a very valuable one (Hawksworth, 1988*a*, 1991).

The names of biotypes, strains, races, mutants, and other genetic variants of an organism will assume increasing importance as information about them

```
ANI    Bactrocera tryoni)p
GEN    Bactrocera
SPE    tryoni
AUT    Froggatt
PUB    1897
FAM    Tephritidae
ORD    Diptera
CLA    Insecta
PNP    p
THS    y
NPS    Chaetodacus tryoni\Dacus tryoni\Strumeta tryoni
CHK    Drew, Hooper & Bateman 1978 p. 45
       \Comb.n. Drew 1989 Mem. Qd Mus. 26 p. 115
OEP    (sunk to Dacus ferrugineus) 4,7-16,19-25,27-37,39,
       42\(as Dacus tryoni) 42,36-48,50-76\(as Bactrocera
       tryoni) 78
BAN    1598
```

Figure 20.1. Sample record from the CABI Arthropod Name Index (ANI) database.

increases. Names play a central role in creating and interpreting information about biodiversity.

An example of the application of information technology to the handling of names is CAB International's new Arthropod Name Index (ANI). The sheer number of insects presents particular problems of nomenclature. ANI contains information about some 70 000 names of arthropods of agricultural significance that have been used in the *Review of Agricultural Entomology* since its inception in 1913. Figure 20.1 shows a sample of one of the 70 000 records in the ANI database, indicating that CAB International's preferred name for the Queensland fruit fly is *Bactrocera tryoni*, and it cites also several non-preferred synonyms (NPS) that have been used for the same organism in the past. ANI makes the connections betwen the different names. By listing volume numbers of the *Review of Agricultural Entomology* in which any of the names have been indexed (OEP), it also provides a key to the extensive literature about the organism. The preferred names from ANI of the more economically important arthropods are listed by Wood (1989).

Codes for organisms

In some contexts, the use of codes instead of names may have advantages. Faust (1979) proposed a 'universal system for coding names of animals, plants and fungi for electronic data processing in crop protection'. Each organism is given a name code of five or six letters, consisting of several letters from its generic name (three for plants; four for animals and fungi) followed by two letters from its specific name. A further systematic code of two or three letters may be added to assign the organism to its systematic position and to remove any ambiguities (Fig. 20.2). Extensive documentation of these

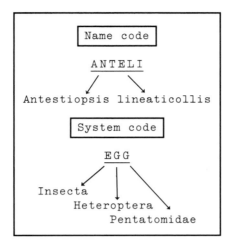

Figure 20.2. Example of 'Bayer code' for an insect (Faust, 1979).

'Bayer codes' has been published by Bayer (1984–86), listing scientific names, synonyms, and vernacular names for some 25 000 organisms. The system allows organisms to be indexed in an unambiguous way that is brief and economical of computer memory. It also has the potential to provide a permanent code for a given organism, whether or not its scientific nomenclature is amended. For full advantage to be taken of these features, the codes need to be widely used and recognized. To date, this has happened only gradually.

Culture collections

The importance of collections of cultures of microorganisms is discussed in other contributions to this book. Such collections represent an archive which, in appropriate contexts, far exceeds in information content the most comprehensive descriptions or illustrations. The archive has an open-ended quality, in that it contains both recognized and as-yet-unrecognized data.

The cataloguing of collections is an important area for the application of information technology. For example, the Microbial Information Network Europe (MINE), was established in 1986 to harmonize and computerize data about some 150 000 strains of microorganisms in European culture collections, and to make them available on-line (Allsop *et al.*, 1989; Hawksworth and Schipper, 1989). The collections include filamentous fungi, yeasts, bacteria, algae, protozoa (an important group, especially in soil, that probably receives disproportionately little attention in the consideration of biodiversity and sustainable agriculture), viruses, and some plasmids and plant and animal cell lines. Data for all the strains are recorded in a standard

```
AATTTGGCTG GTTGTTCGAG CTTAAAGATT TGTT GGGGGC ATGCCCCC CA AGATCTTTTG

AACCGTACAA CAGTTAGGAA ATTTTAATTG AAGGATAAGA TTTGATTTCC TGGAGTATAT

TGTTGATTGA AATTTGAATA TAATATTGAT TGATTGATCC ATGGGTCGTG ATATGAAAAG

TAAAGTTGTG TTGTAAGGGA GGTAATTGAA GTGATCGAAT AATTTGGCTG GTTGTTCGAG

CTTAAAGATT TGTT GGGGGC ATGCCCCC CA AGATCTTTTG AACCGTACAA CAGTTAGGAA

ATTTTAATTG AAGGATAAGA TTTGATTTCC TGGAGTATAT TGTTGATTGA AATTTGAATA

TAATATTGAT TGATTGATCC ATGGGTCGTG ATGTGAAAAG TAAAGTTCTG TTGTAAGGGA

GGTAATTGAA GTGATCGAAT AATTTGGCTG GTTGTTCGAG CTTAAAGATT TGTT GGGGGC

ATGCCCCC CA AGATCTTTTG AACCGTACAA CAGTTAGGAA ATTTTAATTG AAGGATAAGA
```

Figure 20.3. Portion of the nucleotide sequence of the mitochondrial DNA of *Gryllus firmus* (Rand and Harrison, 1989). Boxes mark the start of repeated sequences.

way, which allows for information on, for example, names, biomolecular properties, genotype, growth requirements, enzymes, secondary metabolites, pathogenicity, and so on (Gams *et al.*, 1988; Stalpers *et al.*, 1990).

There are also referral databases such as the Microbial Strain Data Network (MSDN), based in Cambridge, UK, which provides access not to the microbial information itself, but to sources of such information, on a world-wide scale (Kirsop, 1989). A review of the wide range of microbial information services now available is provided by Krichevsky *et al.* (1988).

For certain purposes, an internationally recognized form for coding microbial characters is needed. The Committee on Data for Science and Technology (CODATA) of the International Council of Scientific Unions has sponsored a book (Rogosa *et al.*, 1986) designating standard numeric codes for a wide range of characters. For example, in Section 8, which is about fungal hyphae, the code 8351 means that 'hyphae form thick-walled nodules which interlock to form a plectenchyma', making this character easy to search for in electronic databases, while achieving a substantial saving in computer memory. CODATA have also established an international communications network.

Nucleic acid data

Reference has already been made to the open-endedness of information content in the collections of specimens (p. 249). Another type of information about an organism, its genetic code, specifies it even more completely. Figure 20.3 shows a portion of the nucleotide sequence of the mitochondrial DNA of the cricket *Gryllus firmus*. The boxes mark the start of sequences that are exactly repeated. They were studied by Rand and Harrison (1989)

```
DE    Chironomus thummi thummi globin gene for globin IV
XX
OS    Chironomus thummi (midge, chironome, Muecke)
OC    Eukaryota; Metazoa; Arthropoda; Insecta; Diptera
XX
RL    Nature 310:795-798(1984)
FH    Key        From       To         Description
FH
FT    PRM        228        231        TATA-box
FT    CAP        260        260        cap site
FT    CDS        306        350        signal peptide
FT    CDS        351        758        globin IV
FT    SITE       819        824        polyA signal
FT    POLYA      842        842        polyA site
XX
SQ    Sequence 945 BP;   294 A;   185 C;   160G;   306 T;
      ctttatttat gtggaaattt tttttccaga atatcgagca gaatatcact agtattgaa
      aagaggtaat taaataagct caaattatta tagagtttgt tgaccttttc taatgatta
      gtggttgaaa acagtaaaaa aaacaaaata gaaaatctct tttgattgca taacgatgt
      tcttatctca cagcttttca caataatgtc ttctcaaaat ttttaagtat aaatggagc
      caaatttcga tagtaaatca gttcttcaat tcgtttcaaa gttgtaactt cacaaacca
```

Figure 20.4. Part of a sample record from the Nucleotide Sequence Data Library of the European Molecular Biology Laboratory (Bishop *et al.*, 1987).

because the number and pattern of repeats is itself characteristic of the observed biodiversity within this species.

Computers, like living cells, are designed to manipulate long strings of coded information. Modern molecular biology depends on computer analysis (Lesk, 1988). Laboratories around the world are networked to vast databases of nucleotide sequences, with which sequences under study can be compared (Bishop *et al.*, 1987). Figure 20.4 shows part of an entry from the European Molecular Biology Laboratory (EMBL) Nucleotide Sequence Data Library (Cameron, 1988). A parallel can be noted with the entry from the Arthropod Name Index (Fig. 20.1). The entry starts with the *name* of a midge, includes a bibliographic reference, and then provides data (only part of which is shown) about its globin IV gene.

The use of information technology in the analysis of such data has become commonplace. For example, Figure 20.5 represents a computer display in which two nucleotide sequences are being compared (Bellon, 1988), using a diagrammatic system devised by Gibbs and McIntyre (1970). The two sequences are represented along the sides of a matrix and dots are placed at the intersections of rows and columns containing the same nucleotide or group of nucleotides. This reveals diagonal lines where the two sequences match, and groups of displaced diagonals where they are repeated. The inference of evolutionary relationships from nucleotide sequence data is reviewed in Doolittle (1990).

Analysis of biodiversity needs to encompass biodiversity at the molecular level. Information technology is an essential aid to its compilation and interpretation.

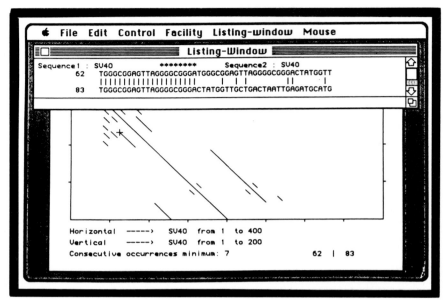

Figure 20.5. Computer display of comparison between two nucleotide sequences (Bellon, 1988). For explanation see text.

Interpretive information

Taxonomic information

Interpreting the diversity of organisms by arranging them in order, usually in a hierarchical manner, is the task of taxonomy. Taxonomy has a special significance in understanding biodiversity by inferring ordered relationships from a mass of unordered detail. Again, information technology has much to contribute. Phenotypic or genotypic information about organisms can be mathematically analysed to infer relationships (Bock, 1988). Many databases have been specially developed to store and interpret taxonomic information (Allkin and Bisby, 1984; Hawksworth, 1988*b*).

The small man in Figure 20.6 holds in his head the concept of a taxonomic reference database that arranges information about all living things in a rational tree of relationships. For the present, this remains a concept only, but the BIOSIS Taxonomic Reference File (TRF) is an attempt to harness information technology to this task (Dadd and Kelly, 1984). To date, the TRF covers the bacteria. It is built to store the taxa of bacteriology, their names, and information about them, and especially to arrange them in hierarchies. Figure 20.7 shows the result of requesting the taxonomic hierarchy above the genus *Streptococcus* or below the family Streptomycetaceae.

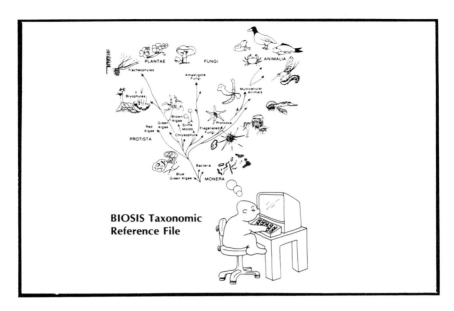

**BIOSIS Taxonomic
Reference File**

Figure 20.6. Diagrammatic concept of the proposed BIOSIS Taxonomic Reference File (TRF). Adapted from Whittaker (1969) and Margulis (1974).

NAME	RANK
Protophyta	Division
Schizomycetes	Class
Eubacteriales	Order
Lactobacillaceae	Family
Streptococcus	Genus

NAME	RANK
Streptomycetaceae	Family
Micromonospora	Genus
Streptomyces	Genus
Thermoactinomyces	Genus

Figure 20.7. Two sample bacteriological taxonomic hierarchies from the BIOSIS Taxonomic Reference File's (TRF) system's BIOSIS Register of Bacterial Nomenclature. The hierarchies are based on those published in *Bergey's Manual* (7th and 8th edns).

```
Enter command: INCLUDE TAXA SAFRICA
203 taxa included.

Use three of the 47 available vegetative characters.

Enter command: CHARACTERS VEGETATIVE

Enter command: 4
4: <habit>
   1. long-rhizomatous
   2. long-stoloniferous
   3. caespitose <Figs 1,7>
   4. decumbent <including 'rooting at the nodes'> <Fig.2>
Enter value: 2
81 taxa remain.

Enter command: 12
12: culms <whether branched above>
    1. branching <vegetatively> above <Fig.2>
    2. unbranched <vegetatively> above <Figs 1,7>
Enter value: 2
68 taxa remain.
```

Figure 20.8. Portion of output from an INTKEY interactive key session to identify a fragment of grass with no floral parts from southern Africa (Watson *et al.*, 1988).

A feature of the TRF is that alternative taxonomic hierarchies can be accommodated, for example those of different editions of *Bergey's Manual of Systematic Bacteriology*.

There are a number of reviews of information systems for taxonomy, such as those by Allkin (1984, 1988), Bisby (1984, 1988), and Pankhurst (1988). Bisby (1989) has reviewed the topic in the context of leguminous plants, to which much attention has been devoted. The International Legume Database and Information System (ILDIS) aims to provide a database of names, taxonomic position, and geographic information for all known legume species. For certain priority species, information on economic importance, botanical features and germplasm resources are to be included. The system uses ALICE software, itself described as a 'biological diversity database system'. A Taxonomic Database Working Group for Plant Sciences, operating under the auspices of the International Union of Biological Sciences (IUBS), has agreed on certain standards for data handling in systems of this sort.

Dallwitz (1980) described a general system for coding taxonomic descriptions, the Description Language for Taxonomy (DELTA), which has been widely adopted as a standard for recording taxonomic data. Software has been developed to produce typeset descriptions from data in DELTA format and, of particular interest in the present context, to provide interactive diagnostic keys (INTKEY) (Watson *et al.*, 1988). INTKEY software guides the user to the identity of an unknown specimen, based only on data that the user specifies as being available. For example (Fig. 20.8), a fragment of a grass with no floral parts is known to come from southern Africa; the system responds that this limits the search to 203 taxa. The user chooses to supply information about plant habit, and then branching of culms, which further limits the possibilities to 81 and then to 68 taxa. Further choices lead closer to a diagnosis. At any point the remaining possibilities can be displayed, and if the user is not satisfied he can choose to return to an earlier point in the search and opt for a different set of characters on which to continue.

Distribution maps

Distribution maps of organisms are another example of interpretive information where information technology can greatly simplify the presentation of data. CAB International's Institutes have for many years prepared distribution maps of plant pathogens and insect pests of plants (*Distribution Maps of Plant Diseases*, 1942 onwards; *Distribution Maps of Pests*, 1951 onwards). For a planned new series covering medical and veterinary pests, a computer system has been developed to plot distributions from the coordinates of locations. The potential to update, and to add extra information on, for example, host distribution, climatic data, frequency, or movement over time, is clear.

Distribution maps are a simple and effective way of displaying information about biodiversity in a form that is easily absorbed. They have special significance for epidemiology, population dynamics and quarantine. The 1989 International Crop Protection Information Workshop (ICPIW) recommended that 'a standard geographical map should be adopted for all pest distribution maps. . . . Such maps need to be validated and kept updated, and pest status for each country should be indicated' (Hopper, 1989).

Multimedia databases

The huge information-carrying capacity of CD-ROM is likely to be the stimulus to the development of multimedia databases that permit interpretation of factual information by creating links between different types of data. CAB International developed a prototype multimedia CD-ROM of crop protection information (Scott and Harris, 1989). A database of names and synonyms of a set of pest organisms provided the links between a large bibliographic dataset, the text of published descriptions of the pests, scanned images of distribution maps, and colour illustrations. This was intended as no more than a prototype. There is enormous further potential to associate different kinds of data through information technology, using databases of names to create links between items in banks of information that may have been constructed for entirely different purposes.

```
IF:

     (1) Time in the season is 1 to 25 DAT
and  (2) Potential rate of BPH population increase is big
and  (3) Immigrant population size is unknown
and  (4) A previous insecticide application has been made
and  (5) The previous spray was applied after 20 DAT
and  (6) Efficacy of previous spray against BPH was good

THEN:

     (1) Future BPH risk medium
and  (2) One spray between 25 and 40 DAT will probably be sufficient.
```

Figure 20.9. Rule from the BPH expert system for management of insecticidal control of rice brown planthopper (Holt and Perfect, 1988).

Decision-making information

Expert systems are computer programs that emulate human decision-making processes based on factual knowledge and sets of logical rules for applying that knowledge to solve problems. Figure 20.9 shows a single rule from an expert system designed to help in the management of the insecticidal control of the rice brown planthopper, *Nilaparvata lugens* (Holt and Perfect, 1988). The IF statement specifies conditions to be satisfied in terms of the values of certain variables (e.g. the value of the variable 'time in the season' may be '1 to 25 DAT'). If all are satisfied, the THEN statement follows: in this instance a moderate insecticidal treatment is suggested.

Expert systems are proliferating rapidly for agricultural applications, including pest diagnosis, pest management, environmental control, irrigation design, financial analysis, extension and advisory services, and many more. Many have a bearing on natural resources, and there is even a journal devoted to expert systems (here called AI for Artificial Intelligence) in this field, *AI Applications in Natural Resource Management*. Lambert and Wood (1989), and Gooch and Cowie (1989), have published partial surveys of expert systems for agriculture and natural resource management. The reviews of Latin *et al.* (1987), Wahl (1989) and Heong (1990) are relevant.

Figure 20.10 shows a sample of output from an expert system that is directly relevant to biodiversity (Messing *et al.*, 1989), NERISK, which assesses the risk to natural enemies of crop pests when pesticides are applied to crops. In this example, the system has assessed the risk to different natural enemy groups (to which estimated importance values have been assigned) of applying bacterial pesticides to apples. The assessments are presented quantitatively in terms of toxicity; further interrogation would reveal bibliographic citations of the major sources on which these estimates are based. Such estimates are reached by the application of IF–THEN rules which aim to emulate the way human judgement might be brought to bear on the problem.

```
RISK ASSESSMENT: BACTERIAL PESTICIDES ON APPLE

Natural Enemy     Importance    Number of     Mean        Variability
                  Value         Records       Toxicity
-----------------------------------------------------------------------
Phytoseiids       5             7             1.0         LOW
Coccinellids      2             17            1.4         MED
Mirids            2             2             1.0         LOW
Parasitoids       1             62            1.6         MED
Chrysopids        1             11            1.4         LOW
-----------------------------------------------------------------------
Weighted average

Note: Mortality to parasitoids is often indirect, occurring when a
      parasitised host dies from the pesticide.
-----------------------------------------------------------------------

ENTER - W: for WHY, X: to EXIT, D: to use DICTIONARY, S: to use
        SCRATCHPAD, R: to leave REMARK for expert, <Carriage Return>:
        to CONTINUE
```

Figure 20.10. Sample of output from NERISK expert system fort assessing risk to natural enemies of crop pests from application of pesticides (Messing, *et al.*, 1989).

Formulation of these rules, as the authors state, presents substantial difficulties because of the limited knowledge base from which they can be derived. In practice, therefore, an expert system may be quite limited in the range of circumstances for which it can make reasonable judgements.

Expert systems are now widely available through personal computers. Lindsey and Novak (1989) provided an analysis of expert systems, including their strengths and weaknesses, and their application to such tasks as interpretation, prediction, diagnosis, design, planning, monitoring, repair, instruction, and control. Compared with human experts they are cheap and consistent, but obviously have limited creativity. One important role for them is in training. A place can certainly be found for them in helping to understand biodiversity.

Predictive information

Many simple or sophisticated mathematical models have been built to predict, for example, the incidence of a plant disease, in which parameters can be varied and the predicted outcome observed; or to make more elaborate predictions such as the rate at which pesticide tolerance may evolve in a population of pathogens. Modelling is a valuable exercise in understanding biological systems, exposing predictions that are initially speculative to the test of comparison with observation. The subject has been reviewed in the context of plant pathology by Campbell and Madden (1990).

Modelling the climate is fraught with problems of scale and of the sheer number of interacting parameters. Predictions of the extent of climatic change are notoriously controversial, and its effects on biodiversity are even

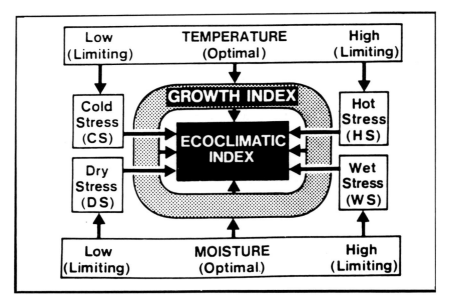

Figure 20.11. Derivation of Ecoclimatic Index for the CLIMEX system (Sutherst and Maywald, 1985). For explanation see text.

more difficult to establish. One example in this field will be discussed. CLIMEX (Sutherst and Maywald, 1985; Maywald and Sutherst, 1990) is a computer system developed in Australia that uses biological or distributional data for an organism to derive an Ecoclimatic Index that is a quantitative estimate of its preferences for growth in terms of temperature and moisture, and the conditions of extreme temperature and moisture that limit its growth. Figure 20.11 illustrates the derivation of the Ecoclimatic Index. Once this has been calculated, the system predicts the potential distribution of the organism from a database of world climatic information (Worner, 1988; Sutherst *et al.*, 1989). CLIMEX may be built into a Geographic Information System (GIS) to integrate climatic factors with other geographic variables (Lessard *et al.*, 1990). Another system developed in Australia, BIOCLIM (Kohlmann *et al.*, 1988), also aims to relate climatic data with the distribution of organisms, using a powerful model for interpolating climatic data between recording stations.

The Queensland fruit fly, *Bactrocera tryoni*, has a distribution limited to eastern Australia, Queensland, and a very few areas to the north. It is the subject of quarantine control to prevent its further spread. Figure 20.12(*a*) shows its potential world distribution as predicted by CLIMEX from its derived Ecoclimatic Index. The largest circles indicate the locations calculated to be most favourable; crosses mark unfavourable locations.

(a)

(b)

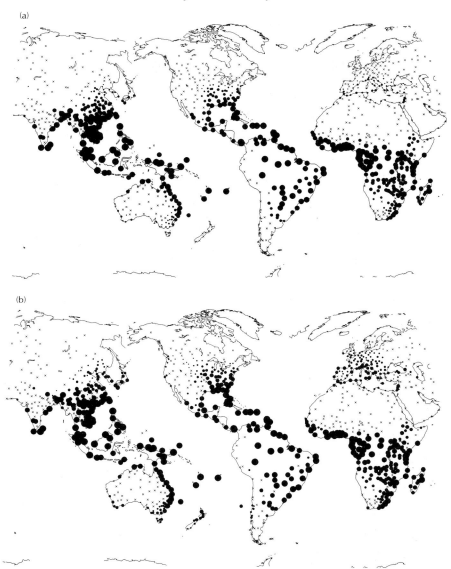

Figure 20.12. (a) Potential world distribution of the Queensland fruit fly (*Bactrocera tryoni*) as predicted by CLIMEX from its Ecoclimatic Index. The largest circles indicate the locations calculated to be most favourable; crosses mark unfavourable locations. (b) Potential distribution assuming a specified climatic change as a result of global warming (0.1°C rise in temperature for each degree of latitude; rainfall 10% less in winter, 20% more in summer) (R.W. Sutherst, personal communication).

Clearly, other factors besides climatic requirements would affect the distribution of the Queensland fruit fly, but, assuming that the climatic model on which CLIMEX is based is reasonably realistic, it appears that quarantine precautions against the spread of the pest are indeed appropriate.

Figure 20.12(b) shows the result of replotting the potential distribution assuming a specified climatic change as a result of global warming (Pittock and Nix, 1986): a 0.1°C rise in temperature for each degree of latitude; rainfall 10% less in winter, 20% more in summer. The prediction is that the populations migrate northwards and southwards into areas that, at present, have a more mediterranean type of climate.

CLIMEX and similar predictive systems can be developed much further, for example to cover ecosystems, rather than one organism at a time, and to handle responses to changes other than climatic ones, though this is a formidable challenge. The result could help to estimate the potential, and perhaps even the future, of biodiversity.

Conclusions

There are grounds for confidence that information and information technology will make an important contribution to our understanding of biodiversity. There will be a need to address information issues specifically, rather than just as an adjunct to biological issues. Information specialists will need to be closely involved with biologists to determine the best strategies for cataloguing, interpreting, and communicating information as a basis for prediction and decision-making.

Standardization of methodology in assessing and recording biodiversity will be important. For example, standardization has been referred to above in the context of biological nomenclature, data coding, taxonomic descriptions, and distribution maps. Standardization also assumes a special significance in information science, to allow the effective communication and networking of information. Networking will assume increasing significance, for applications such as the exchange of molecular or taxonomic data. To illustrate the power of networking, Fig. 20.13 lists the primary user institutions of CGNET, the computer communications network of the CGIAR centres which has revolutionized the transfer of all kinds of agricultural information around the world (Lindsey and Novak, 1989).

The high information content of the study of biodiversity, and its scale, lend themselves to handling by the speed and power of computer processing. Thus, information technology holds an important key to advance in this area. Information technology offers the computer-power to handle enormous databases and to interpret their content; it offers CD-ROM technology to give local access to large arrays of diverse information; it offers expert systems and the challenge of improving their capacity to emulate human

ACIAR	Australian Centre for International Agricultural Research
AIDAB	Australian International Development Assistance Bureau
CGIAR	Consultative Group on International Agricultural Research
CIAT	Centro Internacional de Agricultura Tropical
CICP	Consortium for International Crop Protection
CIDA	Canadian International Development Agency
CIMMYT	International Maize and Wheat Improvement Center
CIP	International Potato Center
CSIRO	Commonwealth Scientific and Industrial Research Organization of Australia
CTA	Technical Center for Agricultural and Rural Co-operation
FRI	Food Research Institute
IBPGR	International Board for Plant Genetic Resources
IBSNAT	International Benchmark Sites Network for Agricultural Testing
IBSRAM	International Board for Soil Research and Management
ICIPE	International Center for Insect Physiology and Ecology
ICLARM	International Center for Living Aquatic Resources Management
ICRAF	International Council for Research in Agroforestry
ICRISAT	International Crops Research Institute for the Semi-Arid Tropics
IDRC	International Development Research Centre
IFDC	International Fertilizer Development Center
IFPRI	International Food Policy Research Institute
IIE	Institute of International Education
IIMI	International Irrigation Management Institute
IITA	International Institute for Tropical Agriculture
ILCA	International Livestock Center for Africa
ILRAD	International Laboratory for Research on Animal Diseases
INTSORMIL	Sorghum/Millet International Research
IPPC	International Plant Protection Center
IRRI	International Rice Research Institute
ISNAR	International Service for National Agricultural Research
NAL	National Agricultural Library
NIFTAL	Nitrogen Fixation by Legumes Center
SOILCON	Soil Conservation Research Unit
TAC	Technical Advisory Committee
UNDP	United Nations Development Programme
USAID	United States Agency for International Development
WARDA	West Africa Rice Development Association
WINROCK	Winrock International

Figure 20.13. Primary user institutions of CGNET, the computer communications network of the CGIAR centres (Lindsay and Novak, 1989).

experts; it offers enormous communication power through computer networks. These tools are waiting to be applied to the information problems of biodiversity.

References

Allkin, R. (1984) Handling taxonomic descriptions by computer. In: Allkin, R. and Bisby, F.A. (eds), *Databases in Systematics*. [Systematics Association Special Volume no. 26.] Academic Press, London, pp. 263–278.

Allkin, R. (1988) Taxonomically intelligent database programs. In: Hawksworth, D.L. (ed.), *Prospects in Systematics*. [Systematics Association Special Volume no. 36.] Clarendon Press, Oxford, pp. 315–331.

Allkin, R. and Bisby, F.A. (eds) (1984) *Databases in Systematics*. [Systematics Association Special Volume no. 26.] Academic Press, London.

Allsopp, D., Hawksworth, D.L. and Platt, R. (1989) The CAB International Mycological Institute Culture Collection Database, Microbial Culture Information Service (MiCIS) and Microbial Information Network Europe (MINE). *International Biodeterioration* 25, 169–174.

Bayer AG (1984–86) *Important Noxious Animals; Diseases of Crops and Useful Plants; Important Crops of the World and their Weeds (scientific and common names, synonyms and computer codes)*. Bayer AG, Leverkusen.

Bellon, B. (1988) Apple Mcintosh programs for nucleic acid and protein sequence analyses. *Nucleic Acids Research* 16, 1837–1846.

Bisby, F.A. (1984) Information services in taxonomy. In: Allkin, R. and Bisby, F.A. (eds), *Databases in Systematics*. [Systematics Association Special Volume no. 26.] Academic Press, London, pp. 17–33.

Bisby, F.A. (1988) Communications in taxonomy. In: Hawksworth, D.L. (ed.), *Prospects in Systematics*. [Systematics Association Special Volume no. 36.] Clarendon Press, Oxford, pp. 277–291.

Bisby, F.A. (1989) Databases, information systems, and legume research. *Monograph of Systematic Botany of the Missouri Botanic Garden* 29, 811–825.

Bishop, M.J., Ginsburg, M., Rawlings, C.J. and Wakeford, R.W. (1987) Molecular sequence databases. In: Bishop, M.J. and Rawlings, C.J. (eds), *Nucleic Acid and Protein Sequence Analysis: a practical approach*. IRL Press, Oxford, pp. 83–113.

Bock, H.H. (1988) *Classification and Related Methods of Data Analysis*. North Holland, Amsterdam.

CAB International (1990) *CAB Thesaurus*. 2 vols. 2nd edn. CAB International, Wallingford.

Cameron, G.C. (1988) The EMBL data library. *Nucleic Acids Research* 16, 1865–1867.

Campbell, C.L. and Madden, L.V. (1990) *Introduction to Plant Disease Epidemiology*. John Wiley, New York.

Dadd, M.N. and Kelly, M.C. (1984) A concept for a machine-readable taxonomic reference file. In: Allkin, R. and Bisby, F.A. (eds), *Databases in Systematics*. [Systematics Association Special Volume no. 26.] Academic Press, London.

Dallwitz, M.J. (1980) A general system for coding taxonomic descriptions. *Taxon* 29, 41–46.

Derwent Publications (1990) *Thesaurus of Agricultural Organisms.* 2 vols. Chapman and Hall, London.

Dextre Clarke, S.G. (1988) The uses and future of bibliographic database systems. In: Hawksworth, D.L. (ed.), *Prospects in Systematics.* [Systematics Association Special Volume no. 36.] Clarendon Press, Oxford, pp. 305–314.

Doolittle, R.F. [ed.] (1990) *Molecular Evolution: Computer Analysis of Protein and Nucleic Acid Sequences.* Academic Press, New York.

Faust, W. (1979) Ein universelles System zur Codierung von Tier-, Pflanzen- und Pilznamen für die elektronische Datenverarbeitung im Pflanzenschutz. *Zeitschrift für Pflanzenkrankheiten und Pflanzenschutz* 86, 396–399.

Gams, W., Hennebert, G.L., Stalpers, J.A., Janssens, D., Schipper, M.A.A., Smith, J., Yarrow, D. and Hawksworth, D.L. (1988) Structuring strain data for storage and retrieval of information on fungi and yeasts in MINE, the Microbial Information Network Europe. *Journal of General Microbiology* 134, 1667–1689.

Gibbs, A.J. and McIntyre, G.A. (1970) The diagram, a method for comparing sequences. Its use with amino acid and nucleotide sequences. *European Journal of Biochemistry* 16, 1–11.

Gooch, P. and Cowie, A. (1989) Survey of electronic databases in crop protection. In: Harris, K.M. and Scott, P.R. (eds), *Crop Protection Information: an international perspective.* CAB International, Wallingford, pp. 237–291.

Hawksworth, D.L. (1988a) Improved stability for biological nomenclature. *Nature,* 334, 301.

Hawksworth, D.L. [ed.] (1988b) *Prospects in Systematics.* [Systematics Association Special Volume no. 36.] Clarendon Press, Oxford.

Hawksworth, D.L. [ed.] (1991) *Improving the Stability of Names: Needs and Options.* [Regnum Vegetabile no. 123.] Königstein: Koeltz Scientific Books.

Hawksworth, D.L. and Schipper, M.A.A. (1989) Criteria for consideration in the accreditation of culture collections participating in MINE, the Microbial Information Network Europe. *MIRCEN Journal of Applied Microbiology and Biotechnology* 5, 277–282.

Heong, K.L. (1990) Computer expert systems for improving insect pest management. *Review of Agricultural Entomology* 78, 1–11.

Holt, J. and Perfect, T.J. (1988) An expert system for insecticide control of brown planthopper (BPH). *International Rice Research Newsletter* 13, 31–32.

Hopper, B.E. (1989) Recommendations of the International Crop Protection Information Workshop. Report of Working Group 3: Plant Quarantine Information. In: Harris, K.M. and Scott, P.R. (eds), *Crop Protection Information: an international perspective.* CAB International, Wallingford, pp. 303–307.

Kirsop, B.E. (1989) ICECC links to the Microbial Strain Data Network. *Information Centre for European Culture Collections, News* 1, 10.

Krichevsky, M.I., Fabricus, B.-O. and Sugawara, H. (1988) Information resources. In: Hawksworth, D.L. and Kirsop, B.E. (eds), *Living Resources for Biotechnology. Filamentous Fungi.* Cambridge University Press, Cambridge, pp. 31–53.

Kohlmann, B., Nix, H. and Shaw, D.D. (1988) Environmental predictions and distributional limits of chromosomal taxa in the Australian grasshopper *Caledia captiva* (F.). *Oecologia* 75, 483–493.

Lambert, D.K. and Wood, T.K. (1989) Partial survey of expert support systems for agriculture and natural resource management. *AI Applications in Natural Resource Management* 3, 41–52.

Latin, R.X., Miles, G.E. and Rettinger, J.C. (1987) Expert systems in plant pathology. *Plant Disease* 71, 866–872.

Lesk, A.M. [ed.] (1988) *Computational Molecular Biology: Sources and Methods for Sequence Analysis.* Oxford University Press, Oxford.

Lessard, P., l'Eplattenier, R., Norval, R.A.I., Perry, B.D., Kundert, K., Dolan, T.T., Croze, H., Walker, J.B. and Irvin, A.D. (1990) Geographical information systems for studying the epidemiology of cattle diseases caused by *Theileria parva. Veterinary Record* 126, 255–262.

Lindsey, G. and Novak, K. (1989) Developments in information dissemination techniques: directions for international agricultural research centres. In: Harris, K.M. and Scott, P.R. (eds), *Crop Protection Information: An International Perspective.* CAB International, Wallingford, pp. 27–48.

Margulis, L. (1974) Five Kingdom classification and the origin and evolution of cells. *Evolutionary Biology* 7, 45–78.

Maywald, G.F. and Sutherst, R.W. (1990) User's guide to CLIMEX, a computer program for comparing climates in ecology, 2nd ed. *CSIRO Division of Entomology Report* 35, 1–38.

McDonald, D. (1989) Recommendations of the International Crop Protection Information Workshop. Report of Working Group 2: Information on Organisms. In: Harris, K.M. and Scott, P.R. (eds), *Crop Protection Information: An International Perspective.* CAB International, Wallingford, pp. 299–303.

Messing, R.H., Croft, B.A. and Currans, K. (1989) Assessing pesticide risk to arthropod natural enemies using expert system technology. *AI Applications in Natural Resource Management* 3, 1–11.

Pankhurst, R.J. (1988) Taxonomic biology seen as a database. In: Bock, H.H. (ed.) *Classification and Related Methods of Data Analysis.* North Holland, Amsterdam, pp. 709–716.

Pittock, A.B. and Nix, H.A. (1986) The effect of changing climate on Australian biomass production – a preliminary study. *Climatic Change* 8, 243–255.

Rand, D.M. and Harrison, R.G. (1989) Molecular population genetics of mtDNA size variation in crickets. *Genetics* 121, 551–569.

Rogosa, M., Krichevsky, M.I. and Colwell, R.R. (1986) *Coding Microbiological Data for Computers.* Springer-Verlag, New York.

Scott, P.R. and Harris, K.M. (1989) Information dissemination techniques and their importance in crop protection. In: Harris, K.M. and Scott, P.R. (eds), *Crop Protection Information, An International Perspective.* CAB International, Wallingford, pp. 13–26.

Stalpers, J.A., Kracht, M., Janssens, D., de Ley, J., van der Toorn, J., Smith, J., Claus, D. and Hippe, H. (1990) Structuring strain data for storage and retrieval of information on bacteria in MINE, the Microbial Information Network Europe. *Systematic and Applied Microbiology* 13, 92–103.

Sutherst, R.W. and Maywald, G.F. (1985) A computerised system for matching climates in ecology. *Agriculture, Ecosystems and Environment* 13, 281–299.

Sutherst, R.W., Spradbery, J.P. and Maywald, G.F. (1989) The potential geographical distribution of the Old World screw-worm fly, *Chrysomya bezziana*. *Medical and Veterinary Entomology* 3, 273–280.

Wahl, V. (1989) Les systèmes-experts de diagnostic en protection des plantes de l'INRA: un point de vue méthodologique. *Phytoprotection* 70, 111–118.

Watson, L., Dallwitz, M.J., Gibbs, A.J. and Pankhurst, R.J. (1988) Automated taxonomic descriptions. In: Hawksworth, D.L. (ed.), *Prospects in Systematics*. [Systematics Association Special Volume no. 36.] Clarendon Press, Oxford, pp. 292–304.

Whittaker, R.H. (1969) New concepts of kingdoms of organisms. *Science* 163, 150–160.

Wood, A.M. (1989) *Insects of Economic Importance: a checklist of preferred names*. CAB International, Wallingford.

Worner, S.P. (1988) Ecoclimatic assessment of potential establishment of exotic pests. *Journal of Economic Entomology* 81, 973–983.

Discussion

Hadley: A recurrent problem in international co-operation is that of ensuring a continuation of activities in a particular field once a given project has ended. A related problem is ensuring that information gained in a country remains in that country once an outside expert has returned home. Information technologies such as those outlined by Dr Scott would seem to hold immense promise in encouraging a continuation of scientific endeavour. What are the opportunities and constraints that need to be addressed for this promise to be realized?

Scott: The opportunities information networking and transfer afford make it much easier for research carried out in one country to be made available in another, for example in the cataloguing of biodiversity and its uses.

Jackson: The information that is lost to developing countries when foreign experts return to their own countries is mainly of two kinds: that which is carried in the head, and that which is described as 'grey literature'. No amount of information technology will make that carried in the head available; it has to be formerly recorded, whether by electronic or traditional methods. With regard to 'grey literature', it would be helpful if organizations who compile databases, such as CAB International, were able to move towards making such literature available electronically, for example through CD-ROM. The CD-ROM is at present a 'read-only' system, and there is as yet no means by which material of local interest can be added to a CD-ROM database at the developing country level. This could change with new technology, but for the time being developing countries must rely on the developed world to make this 'grey literature' available in an electronic form.

Bull: In response to concern expressed on 'downstream' interactions following the intervention of experts in developing countries and their subsequent return to their home institutions, it is my experience that one-way interactions have only limited value and at worst can be counter-productive. The transfer of information, technical skills, and the development of effective programmes are best achieved through two-way working exchanges. Recent experience of the latter mode of interaction at the University of Kent has been through World Bank postgraduate training and a subsequent Overseas Development Administration (ODA) supported co-operative programme 'Biodiversity for Biotechnology' in Indonesia (Chapter 17).

Kishimba: Training should be conducted in the environment where work is to be carried out, for example in tropical forests, savannah, etc. Also, training should be supported by making the equipment required available, such as computers and software.

F. Baker: A UNESCO mission to an African country which had an online information system installed found that the scientists tended to contact colleagues in the north rather than those in the own country or their national information centre; reasons given were relevance of the data and ease of access. I believe there is already a great deal of north–south exchange of information through personal contacts. Dr Scott drew attention to the potential value of computer mapping in relating distributions to global warming; such models could also be of value in predicting the effects of such warming on less-movable organisms.

Arnold: With respect to the constraints of using modern information technology in tropical countries, I am an optimist. I believe there are no insurmountable constraints. Witness, for example, the improvement in telephonic communication in a country such as Niger, where, instead of having to maintain hundreds of miles of telephone wires across the desert, there are now a few strategically placed satellite dishes opening up totally new opportunities in rapid communication. Hardware for networking and CD-ROM is increasingly becoming available at reasonable prices (or through development assistance) in such countries. Such equipment is reasonably robust, easy to use, and in my opinion will revolutionize the ease of communication among scientists and between individuals in the industrialized and developing countries, as well as providing access to previously unavailable databases, expert systems, and other sources of information.

Twenty-one

General Discussion: Session IV

Session Chairman: J.H. Hulse, *1628 Featherstone Drive, Ottawa, Ontario K1H 6P2, Canada*

Session Rapporteur: D.L. Hawksworth, *International Mycological Institute, Ferry Lane, Kew, Surrey TW9 3AF, UK*

Introduction

The principle purpose of this session was to discuss how modern biotechnology might be better employed to understand and employ useful microorganisms and invertebrates. This is a complex, enormously diverse and rapidly growing field as will be evident from the presentations made, but we must be conscious that it has been possible to cover only some of the aspects here.

Discussion

Hawksworth: In this session it has not been possible to have contributions dealing with all areas where microorganism biodiversity is relevant to agriculture and related applications. As a background to this discussion it might be appropriate to identify some of these here: the search for mycorrhizal inoculants for use in commercial forestry in temperate as well as tropical regions; the discovery of endophytic fungi with insecticidal properties inside grasses; the use of additional edible mushrooms and improvement of those strains; single-celled protein production such as the meat-substitute 'Quorn' now available in UK supermarkets; sources of natural pigments, flavourings and fragrances; gut ruminant fungi only now starting to be studied in depth; etc. Further, microorganisms have a tremendous potential role in both monitoring biotic changes in natural habitats we wish to preserve *in situ*, in waste biodegradation, detoxification and utilization, and also in clearing-up after man-made disasters – as in the bioremediation of the 1989 Alaskan oil spill.

The Biodiversity of Microorganisms and Invertebrates: Its Role in Sustainable Agriculture.
Edited by D.L. Hawksworth. © CAB International 1991.

Lynch: We have heard much of the potential to be gained from the exploitation of microorganisms, particularly when genetically engineered organisms are released into the environment. Reintroduction of even indigenous organisms at elevated levels could generate hazards. Antibiotic resistance could develop in the environment because of increased selection pressure. No such problem would occur if those organisms were merely good competitors for the available substrates. Substrate competition appears to be a potentially useful way to proceed with relatively little cause for concern. In view of the threat of antibiosis, such introductions and reintroductions should be monitored as well as those of genetically engineered strains.

Beringer: The UK Advisory Committee on Releases to the Environment does have a role in regulating the release on non-indigenous organisms. It is also possible that controls on the releases of organisms moved around the UK may come in the future. My concern is whether this is realistic or would be acceptable. What should, and what should not, be regulated in this way? Have we really seen problems from the use of the *Bacillus thuringiensis* toxin world-wide that suggest this is needed?

R. Baker: I welcome the proposals made in this session for networking and training, and hope that these will trickle down to developing countries in order to assess low-input systems able to meet the market challenges of the twenty-first century; the cost of production, the quality of the goods, and the quantity produced. I also wonder if we can afford biotechnology in terms of cost and the monitoring of unfamiliar introduced organisms. Networking would enable us at least to access the available knowledge.

Session Chairman's conclusions

Summary

Professor Bull (Chapter 17) described the benefits to be gained from molecular biology research and the many opportunities that can arise from genetic engineering. The knowledge of biodiversity among microorganisms is quite inadequate and many more systematic studies are needed; new genera and species are continually being discovered which have useful properties. Attention was also drawn to the value of looking at mixed symbiotic cultures. He also stressed the need to give emphasis to economic and practical as well as cultural issues. The importance of targeted screening and approaches to antibiosis were highlighted, as was the importance of interdisciplinary approaches, including information science. I understand that the collaborative studies he is involved in in Indonesia have already yielded additional fungi with high lingocellulolytic properties.

Professor Beringer and his colleagues (Chapter 18) considered the methodologies of culture collections, and described the benefits of storing DNA sequences in computers. DNA probes can also be of value in ascertaining the diversity of the biota in soils. Opportunities for genetic engineering include biocontrol, waste disposal, biological ntirogen fixation, etc. They also drew attention to regulatory problems, especially with regard to developing countries. We should perhaps pay more attention to the human factors in our discussions.

Professor Nisbet and Dr Fox (Chapter 19) reviewed the many industrial applications of microorganisms. We have not yet begun to scratch more than the surface of the water or soil for the range of microorganisms from which useful products can be obtained. Biosensors, enzyme inhibition, molecular sizing, were mentioned, and also pharmaceuticals other than antibiotics, promise for some of which lies in the fungal components of lichens.

Dr Scott (Chapter 20) appropriately terminated the session by stressing the need for information: factual, interpretative, predictive information, and modelling (including global warming).

Conclusions

The field of biodiversity in relation to biotechnology is so enormous, that the remarkable potential which exists to be realized for our benefit will not be achieved by independent small groups. There is a need for greater information exchange and co-operation on an international basis. While the research phase in biotechnology is important, from the outset there needs to be a clear concept of how benefits are to be realized and for whom, and what are the resources (human, physical, institutional, organization, and material) needed to implement them. M.G.K. Menon once remarked to the Prime Minister of India that for every rupee put into biotechnology research, 100–200 rupees would be needed for the application of the results. We must not raise unrealistic expectations.

Particular attention should be paid to establishing co-operative research programmes to develop reliable target screening strategies to identify organisms with desirable or useful properties. Also needed are improved systems of communicating information about genetic diversity, and standardized methodologies for determining what is useful in biodiversity.

In order to facilitate the realization of the massive potential that exists among the many groups of organisms, consideration should be given to the establishment of a body similar to the International Board on Plant Genetic Resources (IBPGR) but concerned with microorganisms and invertebrates. That could at least take the form of a central database made available to the whole world, and is particularly important for developing countries, many of which do not have access to the information they need. CD-ROM offers particular opportunities waiting to be realized.

There is a need for a greater understanding of both natural ecosystems and agro-ecosystems, and a preference for using wild-type species rather than genetically engineered strains. Genetically engineered strains were seen as an often useful intermediary step in identifying useful naturally occurring organisms.

Greater systematic protection of natural ecosystems was also strongly supported as that is the source of the biodiversity being sought.

Biotechnology must have a clear definition of purpose, particularly who is intended to benefit, how that benefit is expected to be implemented, and how the resources needed are to be provided. The emphasis needs to be on the functional diversity relevant to biotechnology development, that is on functional properties rather than simply scientific interest. In emphasizing the human dimension, training programmes, particularly in developing countries, are to be strongly supported.

As it will be difficult for small countries to exercise adequate monitoring or control of the release of large numbers of genetically engineered or other organisms, regional organizations need to be encouraged to address this problem.

More systematic research is needed to understand and enlarge the potential of microbial diversity to provide the many novel substances that can be used to treat the maladies of humans, animals, and crop plants. Research into isolation methods, culture conditions, and preservation systems are also necessary to realize the potential of microbial diversity. Communications networks set up to exchange data need to embrace the industrial sector as well as academic and government institutions; international co-ordination is required.

Co-ordination and integration must be interdisciplinary, and not left to a single discipline. The multifarious scientific interests cover mycology, microbiology, entomology, agronomy, chemistry, biochemistry, and information science.

There is also a need to compare the microbial diversity of tropical *versus* temperate ecosystems, with a particular focus on appropriate indicator organisms. DNA probe techniques have great potential in facilitating the study of the occurrence of all types of organisms in soil and water. Greater use can be made of gene banks and DNA databases as a repository of knowledge on genetic variation.

We also need to use the best, and also to develop improved, methods for the long-term storage of microorganisms in culture collections. Screening programmes for novel organisms should be devised rather than only seeking to use gene manipulation techniques. This should serve to expand the world's culture collections, and to foster the recognition that existing diversity is extremely important and can be profitably utilized.

Conclusions

Twenty-two

Chairman's Remarks

M.S. Swaminathan, *M.S. Swaminathan Research Foundation, 14, 11 Main Road, Kottur Gardens, Kotturpuram, Madras 600 085, India*

ABSTRACT: The following 10 points are considered: (i) sustainable agriculture: need for conceptual clarity; (ii) sustainable agriculture matrix; (iii) biological diversity; (iv) biosystematics; (v) priorities in conservation; (vi) ecological economics; (vii) technology blending; (viii) biotechnology; (ix) biological diversity and global warming; and (x) where do we go from here?

Introduction

The Chairmen of the four different sessions have already presented their conclusions; so I will not go over that ground again. In the course of listening to them, and listening to the presentations, I have noted 10 points on which I would like to share with you my thoughts and views.

Sustainable agriculture: Need for conceptual clarity

This workshop was called 'The Ecological Foundations of Sustainable Agriculture', with special reference to biological diversity in microorganisms and invertebrates. However, there was only one comment, by Dr F. Baker on the definition of 'sustainable agriculture'. He recalled the definition proposed by the Technical Advisory Committee (TAC) to the Consultative Group on International Agricultural Research (CGIAR). Although it is easy to define sustainable agriculture in the way that TAC has done, it is difficult to articulate the pathways by which sustainability can be imparted to a production system. In the field, we are confronted with problems

The Biodiversity of Microorganisms and Invertebrates: Its Role in Sustainable Agriculture.
Edited by D.L. Hawksworth. © CAB International 1991.

of opposing requirements. Shifting cultivation is often referred to as one of the reasons for deforestation today, but at the time shifting cultivation was originally developed as a method of soil fertility restoration it was a very sustainable system. Shifting cultivation became unsustainable only when the pressure of population on land grew.

Another example is provided by the 'green revolution' technology based on high yielding strains of wheat and rice. To some, the green revolution is an ecological disaster. But according to the United Nations Environment Programme (UNEP), a large percentage of the loss of forests comes from the expansion of agriculture. In fact, UNEP and the Food and Agriculture Organization (FAO) have calculated that over 12 million hectares of forest are being cleared each year for the cultivation of annual crops. What will happen if our population goes on increasing and our yields remain almost stagnant? Obviously, more forested land will be cleared. The green revolution technology, based on better yielding genetic strains coupled with good management, is best described as land-saving agriculture. Before the advent of high yielding varieties of wheat in 1965, farmers in India harvested about 12 million tonnes of wheat from 14 million hectares. In 1989, Indian farmers harvested over 55 million tonnes of wheat from about 23 million hectares. To produce 55 million tonnes of wheat at the 1965 yield level we would have required over 55 million hectares of land. Therefore, when we talk about sustainability, one has to look at the totality of the picture. The challenge lies in developing 'green' technologies and public policies which can help to promote sustainable advances in biological productivity.

In this context, the present Workshop should be regarded as the first in a series of workshops needed to address different aspects of the Ecological Foundations of Sustainable Agriculture. We must look at different components of farming in greater depth, as, for example, the whole question of yield improvement in relation to sustainability. What are the pathways to achieving what is now called low input sustainable agriculture (LISA)? At this Workshop, we have discussed the scope and limitations of both biopesticides and biofertilizers. We did not discuss very much the role of green manure crops. At a joint symposium of the Committee on the Application of Science to Agriculture, Forestry and Aquaculture (CASAFA) and the International Rice Research Institute (IRRI) organized at IRRI in 1986, the role of green manure crops in sustainable agriculture was considered at length; the proceedings have been published.

The term 'low input agriculture' sometimes creates the impression that inputs are not needed for output. It is important to emphasize that what LISA is concerned with is not the quantity of nutrients required by the crop to produce a certain quantum of output, but their source. Low input agriculture implies less dependence on market-purchased chemicals. For example, a 5 tonne rice crop would need a minimum of 100 kg nitrogen. This can be produced partly through the cyanobacteria in *Azolla*, green manure, and organic recycling, and partly through mineral fertilizers. Such a system of integrated nutrient supply, coupled with an integrated pest management system, will reduce dependence on mineral fertilizers and chemical pesticides. Unless there is conceptual clarity, the sustainable agriculture movement can give a set-back to efforts in the area of productivity improvement. China has already only 0.1 ha capita^{-1} of arable land. Therefore China and India can feed their growing populations only through a vertical improvement in productivity. Technologies for sustainable agriculture should protect both global food security and the livelihood security of the poor. We need for this purpose new measurement tools, as well as a well-defined set of indicators of unsustainability. This is already taking place. For example, environmental economists are working on methodologies for integrating the principle of sustainability in economic growth. The concept of net domestic product (NDP), in contrast to the gross domestic product (GDP), is being refined. The World Resources Institute (WRI) has used these concepts in studies in Indonesia and elsewhere. NDP gives an idea of real advances in national income as it takes into account the depreciation or damage caused to environmental capital stocks.

Agronomists are also developing methodologies for measuring yield from the point of view of ecological sustainability. For example, at a meeting I organized in 1987 at the International Rice Research Institute (IRRI), it was agreed that the following formula should be used to express yield.

$$\text{Yield} = \frac{\text{Output value}}{\text{Input value}} + \text{impact on environmental capital stocks}$$

To use this formula in practice, reliable methods of measuring the impact of productivity on soil health, groundwater status, genetic diversity, and release of methane, nitrous oxide, etc., will have to be developed. Workers in biological diversity will also have to develop their own terminology, their own methodology of measurement of biological diversity. It is easy to compile information for *Red Data Books*, but it is more difficult to organize measures to save the species under threat of extinction. The discussions at this Workshop have revealed that we do not know how many species exist on our planet. An estimate of 80 million species was mentioned taking into account undescribed species of invertebrates and microorganisms (Hawksworth and Mound, Chapter 3).

It is obviously difficult to be precise in describing what we do not know. This underlines the importance of reviving the interest in taxonomy. There is also a need for more intensive efforts in saving the species listed in *Red Data Books*, and in 'protecting the protected areas', as often in developing countries even protected areas are under threat from anthropogenic pressures.

Sustainable agriculture matrix

Because of its complexity and feedback linkages with numerous parameters, there is a need for a matrix approach for achieving our goals. A sustainable agriculture matrix on one axis has three groups of parameters, that is technology, services, and public policies. On the other axis, we have parameters relating to ecology, economics, and equity. How can we ensure that technologies, services, and public policies, satisfy the needs of ecology, economics, and equity?

Obviously this task is a complex one. It needs a systems approach. Both multi-disciplinary and multi-institutional collaboration will be needed, if the needs for convergence and synergy between technology and public policy are to be satisfied. We should always emphasize that inputs are needed for output, and that inputs can be measured in terms of energy, whether for ploughing, or pumping water, or for the supply of nutrients. There is a need for developing a sustainable agriculture matrix for each major biogeographical zone or farming system.

Biological diversity

My next point is the question of biological diversity with particular reference to invertebrates and microorganisms. We had questions about the quantitative and qualitative dimensions of diversity. Obviously both will have to be taken into consideration. There are several estimates about the number of species on our planet ranging from 10 to 80 million. Obviously our ignorance in this area is great.

Nevertheless, what we do not know should provide impetus for our quest to understand the quantitative and qualitative dimensions of diversity. We must continue our studies and refine our estimates of what is there, at the three levels of measurement used by the World Conservation Monitoring Centre at Cambridge: ecosystems, species, and intraspecific diversity. At the intraspecific level, both metric (polygenic) and non-metric characters are involved. Universally accepted measurement tools will have to be developed. The sustainable management of biological diversity at the ecosystem level is creating increasing conflicts between proponents of development and of conservation. Even in a country such as the USA, I found one particular Gallup poll that showed that three out of every four persons interviewed were environmentally literate and wanted to be considered as belonging to the category of 'greens'; there is now an ongoing controversy on what is worth saving, the spotted owl or 30 000 jobs. If this is so in one of the richest countries in the world, you can see the dilemma in poor nations. Controversies as to what is the more important, 1 000 megawatts of power or 100 000 hectares of forest, are frequent. Such controversies will always persist and it is difficult to resolve them to the entire satisfaction of all concerned. However, in the ultimate analysis, unless every country achieves harmony between human and animal populations and the natural resource endowments of the country, and fosters sustainable life-styles, it will be very difficult to resolve these conflicts. Therefore, I think we should consider biological diversity in human terms and not just in its taxonomic and genetic dimensions. Dr J.D. Holloway (p. 61) gave an example on how conservation measures which often threaten the livelihoods of the inhabitants of forests can be used to strengthen their livelihood security. He gave the example of the economic value of insect specimens.

Unless conservation measures strengthen the livelihood of the people, rather than erode their livelihood, security of conservation of unique ecosystems can become a lost cause in poor and population-rich countries. This complex relationship between human beings and the conservation of biological diversity deserves greater attention from social anthropologists working in close collaboration with biological scientists. This is essential for ensuring the effectiveness of *in situ* conservation of biological diversity in the form of biosphere reserves, national parks, and protected areas.

Biosystematics

Interest in the discipline of systematics is unfortunately on the decline among young scholars. Professor D.J. Greenland rightly said to me that the taxonomist should be included in the *Red Data Book* as a vanishing tribe. This situation comes partly from the lack of social prestige to this kind of work, and second from inadequate opportunities for professional growth, promotion, and so on. Taxonomic positions become a blind alley in many organizations. Hence, we should view this problem not only in terms of training, but in terms of subsequent professional growth, and the infrastructure needed for modern taxonomic research, including opportunities to use DNA probes and DNA fingerprinting techniques.

Fortunately, a number of bilateral donors today also include the provision of some basic equipment along with the training programmes. When I was in IRRI, for example, we had a computer simulation model training programme, jointly developed with scientists in Wageningen. An interdisciplinary group of research scholars was invited from each country participating in the programme, and each group was given a personal computer at the end of the training programme. Back home they performed some experiments, and came back after 6 months to another workshop to report on the results of the studies they had carried out. Training should not just be looked upon in terms of numbers, but in terms of post-training performance. Dr R. Gamez made a very important point when he stressed the importance of a partnership between CAB International and local organizations. His own organization, the National Biodiversity Institute in Costa Rica, is a good example of how to derive maximum benefit from external expertise (p. 64). Dr R. Gamez has wisely built up a critical mass of expertise within the country in taxonomic identification. He has emphasized the importance of involving local communities in identification. In forest areas women, for example, may be totally illiterate but they know the name of every plant in the forest in the local language. They may not know the botanical name, but they have a local name for every plant and they know their uses. When you are bitten by a scorpion or a snake, they will tell you which plant extract is of use as an antidote. This enormous local wisdom has unfortunately not been chronicled and is becoming a dying wisdom. I hope Dr Gamez will chronicle the innovative work he is doing to harness local knowledge and experience in the description and classification of our flora and fauna.

Community involvement in the conservation of biodiversity becomes easier when the community itself enjoys, particularly the younger people, knowing what the plants are. It is a huge task. In May 1990, on behalf of the Commonwealth Secretariat, I led a team to Guyana, whose President had offered to the Commonwealth 400 000 hectares (1 million acres) of tropical rainforest for a Commonwealth project to demonstrate how a rainforest can be managed in an ecologically sustainable manner. Nowadays, we add easily prefix 'sustainable' before every development activity. The President of Guyana has thrown out a challenge, to demonstrate how 1 million acres of prime rainforest can be managed in a sustainable manner. Therefore, when we went into this beautiful forest in its pristine purity, we discovered the dimensions of the taxonomic challenge before us. A few taxonomists from the Netherlands had identified about 1 500 species in that area. There has been very little work on invertebrates and microorganisms. A huge task awaits us in getting the flora and fauna, including the soil microflora and fauna, identified and catalogued.

Systematics and taxonomy are increasingly using a wide array of tools, including molecular biology. At the same time, we should again revive interest among young scholars interested in field identification and description. This will be possible if personnel policies enable such people to progress in terms of financial and professional advancement.

Priorities in conservation

Recent advances in genetic engineering, making the transfer of genes across sexual barriers possible, have made all genetic variability valuable. Hence, the determination of priorities in conservation is difficult. Nevertheless, habitat conservation and the protection of ecological and economic key species deserves high priority. Special efforts are needed to strengthen ongoing efforts in the *ex situ* conservation of diversity in invertebrates and microorganisms. Sampling techniques are important in capturing a representative sample of existing biodiversity for *ex situ* conservation. For example, Sir Otto Frankel, a pioneer in the conservation of crop plant genetic resources, feels that many gene banks have too many samples.

What is important, therefore, is the conservation of habitats. There are divergent viewpoints in this matter. Among experts there must be some agreement on what to conserve, how to conserve, and the kind of sampling procedures we should adopt. Particular mention was made of parasites and predators and their original homes (Waage, Chapter 13). I think there is a great scope here. I am glad this year's World Food Prize is being awarded to Dr John Niederhauser, who about 40 years ago established a 'hot spot' screening nursery in the Toluca Valley near Mexico City for helping scientists all over the world screen potato material for resistance to a broad spectrum of races of *Phytophthora infestans*, the fungus causing the late blight disease of potatoes. We need many such hot spot screening locations and the conservation of such spots should also receive priority.

Ecological economics

Those in charge of allocating funds for the conservation of genetic diversity often demand data on the economic benefits flowing to the community as a result of genetic conservation. The example of the rice strain IR 36, which has genes for resistance to brown planthopper from *Oryza nivara*, is often cited to underline the economic significance of genes for resistance to biotic and abiotic stresses occurring in wild species. In this connection the pedigrees of modern varieties are very interesting; most of them have in their parentage many land races. The new varieties in many cases are developed from a very broad spectrum of genetic material.

The economic value of genetic variability in microorganisms, such as those yielding antibiotics or toxins like the Bt toxin (*Bacillus thuringiensis*; Waage, Chapter 13) is well-known. The role of soil microorganisms in building the biological potential of the soil is also well-known, and the value of biofilters is also being recognized more widely. Indeed, one enterprising young man in India has set up a large earthworm factory producing several tonnes of earthworms for sale each year. It is important that soil scientists look at biological methods of soil fertility restoration, as in developing countries there are large areas of degraded lands where the biological potential needs upgrading.

It is also necessary to have simple and locally understandable methods of measurement of the biological potential of the soil. We define desertification as a human-induced process which either diminishes or destroys the biological potential of the soil. If simple measurement tools can be developed, it would become easy for the local people to understand how this process works. In the days of British rule in India there was a simple method of expressing yield. In the early part of this century, the Indian rupee had 16 annas. An 'annavari' system of expressing the impact of national calamities like drought and floods on crop yields was developed; for example, a 4-anna crop meant 25% of the normal yield. Detailed administrative measures were prescribed for coping with situations where the yield drops to below 75% of normal. We need similar methods of expressing the biological potential of the soil.

Technology blending

There is no other pathway open to developing countries for achieving food security other than by sustainable advances in productivity, whether animal or crop productivity. At the same time, this has to be achieved without detriment to the long-term production potential of the soil, water, and other natural assets. Therein lies the challenge, that is the challenge of the new technology of sustainable agriculture. In my view this should integrate the best in biological technology, space technology, information technology, and management technology, with traditional practices and technologies. The blending of traditional and frontier technologies is a new science of technology development which I think offers a great deal of opportunity. As management holds the key to optimizing the benefits from the available natural, financial, and human resources, sustainable management procedures deserve particular attention.

Biotechnology

We have discussed the biosafety aspects of biotechnology research (Beringer et al., Chapter 18). There is a fear in developing countries that experiments not permitted in developed countries may be performed in their countries. Therefore, the stress on an international code of conduct adopting a United Nations system is a good one.

Another aspect which is increasingly causing concern among nutritionists in particular, is the methodology of wholesomeness assessment. The bovine growth hormone is one aspect which is very much in the news because European countries are not allowing the import of beef or milk from cattle raised with bovine somatotropin. Several states in the USA have also prohibited the sale. In the case of food irradiation, as many of you may recall after world War II, what atomic scientists thought would be a great boon to food technology, soon proved to be an area requiring careful scientific studies. The criteria for assessing food safety soon included genetic parameters. Hence, food safety issues in relation to products arising from biotechnology research should receive attention from the early stages of such research. Genetic engineers should, right from the beginning, collaborate with nutritionists, and home scientists, to develop an agreed methodology of assessment.

Gene patenting is a very complicated issue and in the GATT-TRIPS (Trade Related Intellectual Property Rights) negotiations, there is a serious north–south divide on this issue. This controversy has grown following the expansion of plant breeders' rights in developed countries. At the moment, there is a UNEP sponsored group working on the development of an international convention on biological diversity. They are hoping by June 1992, when the UN Conference on Environment and Development will be held in Brazil, that some kind of international agreement can be developed which will help to save and share biodiversity as a common human heritage. These are areas which are now under discussion.

I can only hope that the knowledge generated at this workshop on a neglected aspect of biodiversity, namely the genetic diversity in microorganisms and invertebrates, will also be used while finalizing the text of the International Biodiversity Convention. Industrialized nations will have to strike a balance between considerations of planet and patent protection. There cannot be a better common future for human kind without a better common present. The World Bank, in its new development report, says that over 1 billion people are now living below the poverty line, out of the 5 billion on planet earth.

Biological diversity and global warming

Both Scott (Chapter 20) and Stewart (Chapter 1) referred to this issue. Professor Stewart reminded us that some algae have a very high tolerance to high temperatures. At my centre in Madras, we are establishing a Genetic Resources Centre for adaptation to sea-level rise. The aim is to conserve species of mangroves, sea grasses and rice strains which can withstand seawater intrusion. Coastal ecosystems are being denuded, either for tourism or brackish water aquaculture, or for other purposes. In the Philippines I have seen considerable coastal mangrove forests being replaced by aquaculture ponds. Aquaculture is important, but the mangroves are equally important to withstand damage from coastal storms. This stresses the need to save genes which may help to both avoid and adapt to new weather patterns. Anticipatory research needs greater support. We also need more well-planned work on the impact of ultraviolet B radiation on crop and animal productivity.

Where do we go from here?

I hope this Workshop will be the first in a series dealing with the Ecological Foundations of Sustainable Agriculture. We have dealt with, in a somewhat broad sweep, invertebrates and microorganisms, but I think we should take up the important issues raised here and examine them in greater depth.

An organization like CAB International, which is a veritable mine of valuable information and expertise, can do a great deal to promote the sustainable agriculture movement. Second, in terms of training, many speakers have emphasized the kinds of training to be given. But as far as I can see, taxonomy needs priority attention, and I hope that biotechnology companies will put some money there. It is in their own interests to do so. If this happens, we will need large numbers of trained people. CAB International will have to promote a consortium of training institutions, some kind of biosystematics network. A global network in biosystematics will also help to provide this science with the necessary financial support and scientific and social prestige. Third, I would like to suggest that CAB International within itself should have a standing working group on biological diversity, because it has four major institutes with special expertise in the area. Finally, publication, communication and information are strong points of CAB International. CAB International could, hence, examine how it can help in spreading information on the ecological foundations of sustainable agriculture.

Acknowledgements

May I thank and congratulate Professor D.J. Greenland, Mr Don Mentz, and the various CAB International scientists for organizing this most timely and meaningful Workshop.

Twenty-three

Statement of Findings

Introduction

A Working Group was convened under the chairmanship of Professor D.J. Greenland in the rooms of the International Institute of Entomology, on the day immediately following the Workshop, 28 July 1990, in order to compile the findings of the Workshop based on the papers presented and discussions.

The Group comprised: Dr M.H. Arnold (Technical Advisory Committee, CGIAR; Vice-Chairman CASAFA), Dr F.W.G. Baker (Scientific Secretary, CASAFA), Dr R.J. Baker (CAB International Liaison Officer, Jamaica), Mr A. Bennett (Chief Natural Resources Advisor, ODA), Dr J. Boussienguet (National Biological Control Programme, Gabon), Mr R. Dellere (Head, Technical Division, CTA), Dr L.E. Elliott (USDA), Sir Leslie Fowden (Lawes Trust Senior Fellow, UK), Professor D.J. Greenland (Director Scientific Services, CAB International), Dr K.M. Harris (Director, International Institute of Entomology), Professor D.L. Hawksworth (Director, International Mycological Institute), Dr J.D. Holloway (International Institute of Entomology), Dr J.M. Hulse (Chairman, CASAFA), Mr A.C. Jackson (Technical Advisor, CTA), Dr R. Lal (Professor of Soil Physics, Ohio State University), Dr J.K. Leslie (Executive Director, Division of Plant Industry, Queensland), Professor J.M. Lynch (Theme Co-ordinator, OECD, Head, Microbiology Department, Horticultural Research Institute, Littlehampton), Mr K.H. Mohamed (Regional Representative, CAB International, Malaysia), Mr D. Moses (Regional Representative, CAB International, Caribbean and Latin America), Professor L.A. Mound (Keeper of Entomology, The Natural History Museum), Dr R.L.J. Muller (Director, International Institute of Parasitology), Mr T.J. Perfect (Deputy Director, Natural Resources Institute, Chatham), Dr G.H.L. Rothschild (Director, Australian Centre for

The Biodiversity of Microorganisms and Invertebrates: Its Role in Sustainable Agriculture.
Edited by D.L. Hawksworth. © CAB International 1991.

International Agricultural Research, ACIAR, Canberra), Dr S.S. Sastroutomo (Asian Plant Quarantine and Training Institute, PLANTI, Selangor), Dr P.R. Scott (Head, Division of Crop Protection and Genetics, CAB International), Professor I. Szabolcs (Deputy Secretary-General, Research Institute for Soil Science and Agricultural Chemistry, Hungarian Academy of Sciences), Dr J.K. Waage (Deputy Director, International Institute for Biological Control), Dr M. Williams (Acting Executive Director, Bureau of Rural Resources, Canberra).

Statement of findings

There is a well recognized need to maintain a stable base for agricultural production which is increasingly jeopardized by the world-wide changes in land use, population growth and its demands on the environment, and the threat of climatic change. The range of diversity contained among the fauna and flora, and certain special habitats, are at particular risk.

Much attention has been focused on the loss of biodiversity among vertebrates and higher plants, where changes in habitat produce easily observed losses in the diversity of species present. Actual losses in numbers of species are likely to be much greater among the less readily observed invertebrates and microorganisms living in the soil and within the forest canopy. These organisms, while less ornamental, play an exceedingly important role in maintaining and promoting soil fertility by recycling the nutrients contained in decaying organic matter, by fixing nitrogen from the air, and by detoxifying pesticides and other waste products. They are also an essential source of many valuable products, such as antibiotics and other pharmaceuticals, and have an important role in most biotechnological and bioindustrial processes.

The Workshop reviewed the importance of biodiversity among invertebrates and microorganisms, what is involved, why it is useful and how it can be preserved and better used. The following summarizes the main findings which should be of interest to scientists and policy makers alike:

1. The needs of human populations for food, fuel and fibre have historically been supported, directly or indirectly by genetic diversity among the microorganisms and invertebrates. These organisms perform functions which prime and fuel the metabolism of soils, plants and animals. The development of sustainable agro-ecosystems and the enhancement of agricultural productivity will depend increasingly on the maintenance of such diversity for:

(a) The improvement of soil structure and fertility, through the decomposition of organic material added to the soil, and the detoxification of pesticides and other pollutants.

(b) The provision of biological controls for insect pests and diseases of plants and animals.

(c) Processing and enhancing the nutritional value of foods.

(d) Maximizing the potential for novel products for use in the pharmaceutical and other industries.

2. We need to increase our knowledge of the nature, extent, functions and potential usefulness of the genetic resources to be found in microorganisms and invertebrates and how to protect different ecosystems as reservoirs of such biodiversity.

3. To do this will require that:

(a) Existing genetic potential in invertebrates and microorganisms is conserved by the preservation of natural and man-made systems and sites, and, where there are difficulties in recovering microorganisms from the environment, by maintaining culture collections of organisms of current and potential value*.

(b) Research is undertaken on the measurement of the diversity of organisms, and of the gene pool contained, and to identify the ecological interactions and the role played by different organisms in agricultural and natural systems. This is particularly pertinent for pest management and, for example, in the role of soil organisms involved in the restoration of degraded lands, the decomposition of carbon compounds and the evolution of carbon dioxide and other greenhouse gases. The likely effects of climatic change on the activities and conservation of invertebrates and microorganisms, their migration and the development of new associations are, as yet, almost wholly undeciphered. While their involvement in the release of carbon dioxide and other greenhouse gases is well established, much remains to be learnt of the factors controlling the quantities and rates of release, and of the organisms involved.

4. Knowledge of the functional role of many microorganisms and invertebrates in agricultural and other ecosystems is inadequate. Limited understanding of their role in ensuring the continued stability of ecosystems requires that much more research be conducted to obtain the information which will enable the importance of biodiversity to be assessed and managed so as to promote the sustainability of agricultural production.

5. The promotion of biodiversity alone will not eliminate the need for improvements in production and husbandry measures. It will, however, contribute to the development of diverse systems able to sustain a low level of production on marginal lands, while making more sustainable the robust,

---

* The Workshop noted the work of IUCN – The World Conservation Union – and World Federation for Culture Collections (WFCC) on these topics, and stressed the need to consider microorganisms and invertebrates more fully in the selection of reserves.

high production systems of more fertile areas, and minimizing the need for external inputs.

6. Habitat disturbance, pollution and climate change all cause changes in the biological population, particularly in ecologically sensitive areas. Invertebrates and microorganisms need to be monitored to follow the effects of these changes. They can be used as indicators of the stress applied to the environment.

7. Current capabilities for the recognition, characterization and study of microorganisms and invertebrates are insufficient to meet current needs. The need to support effective conservation and utilization of biodiversity requires development of skills in biosystematics and related disciplines world-wide.

The shortage of biosystematists can be accommodated in the short term only by the fullest collaboration between centres of expertise. This may best be done through the establishment of a formal Network of Biosystematic Centres. This will involve strengthening of existing centres of activity, and the development of new centres through training, infrastructure building and information transfer.

8. The benefits for mankind of retaining a wealth and diversity of biological resources should be more clearly recognized. There is particular need to promote an understanding of the potential of invertebrates and microorganisms for advancing human welfare. It is known that improved management of land and other resources, through the imaginative use of biodiversity, can increase the efficiency of agricultural production and so improve social and economic welfare. Soil organisms have provided a wealth of valuable products such as antibiotics, and much remains to be learnt of the value of the products which may be obtained from the vast biological resources of the natural world.

9. Education and training must reflect the need to increase awareness of the significance of biodiversity in agriculture and the environment, including the problems of protecting ecologically vulnerable areas. This may be achieved by regional and national policies for primary and secondary education. The media and local natural history and conservation societies also have an important role in complementing and augmenting such policies, and may help to stimulate their initiation.

10. Information systems are essential for recording, interpreting and making available the accumulated knowledge of biodiversity. The specific needs include:

(a) Research on the most appropriate ways of recording, collecting and interpreting data and for disseminating information through participatory networks.

(b) Standardizing the methodologies for the collection and recording of information.

(c) Co-ordinating and implementing the acquisition of information.

(d) Co-ordinating and implementing the dissemination and networking of information.

The standards being developed by the International Union of Biological Sciences (IUBS) through its Taxonomic Databases Working Group may be particularly valuable in relation to these needs.

11. Biodiversity covers immense and complex issues in which many agencies and institutions are active world-wide. There exists a need for an international mechanism to co-ordinate, analyse and disseminate the relevant information. This will include:

(a) The standardization of scientific and information methodologies.

(b) The processing of information on the significant hazards to the protection of biodiversity.

(c) The monitoring of world-wide activity relevant to biodiversity and its application.

(d) Promoting protocols for the control of the release of exotic unfamiliar organisms.

(e) Recommending courses of action to ensure the protection of indigenous property rights and assurance of fairness for those whose survival, social and economic welfare depend on the conservation and prudent employment of natural resources.

(f) Ensuring that resources are directed to enable developing countries to implement an optimum course for the sustainable utilization of their natural resources.

12. It is hoped that these findings will be considered for inclusion in the Agenda of the United Nations' Conference on the Environment and Development in 1992, and other appropriate initiatives.

Index